THE COMMISSION OF THE EUROPEAN COMMUNITIES

RELIABILITY OF SUB-SEABED DISPOSAL OPERATIONS FOR HIGH LEVEL WASTE

This report was prepared for the European Atomic Energy Community's Cost-sharing Research Programme on "Radioactive Waste Management and Disposal".
Under contract No 395-83-7 WAS UK
Project 7B: "Disposal in Sea-bed Geological Formations".

RELIABILITY OF SUB-SEABED DISPOSAL OPERATIONS FOR HIGH LEVEL WASTE

by

M. M. SARSHAR

Taylor Woodrow Construction Limited
Southall, Middlesex, UK

Published by GRAHAM & TROTMAN Ltd
for the Commission of the European Communities

Published in 1986 by Graham & Trotman Ltd
66 Wilton Road, London SW1V 1DE, UK
for the Commission of the European Communities,
Directorate-General Information Market and Innovation,
Luxembourg

EUR 10542

ISBN 0 86010 835 x

LCCCN and British Library CIP data available from publisher.

Printed and bound in Great Britain

CONTENTS

LIST OF TABLES

LIST OF FIGURES

LIST OF APPENDIXES

SUMMARY

This report gives a summary of the reliability study related to the disposal of high level waste in deep ocean sediments. Two methods of disposal, by drilled emplacement and by penetrator as developed by Taylor Woodrow and Ove Arup & Partners respectively, have been assessed in this study.

For assessment of the reliability, two well established methods have been used. These are:

a) Failure mode effects and criticality analysis (FMECA)
b) Fault tree analysis (FTA).

For the fault tree analysis, two major undesired events have been considered. The events are:

i) Loss of lives of personnel involved in the operation
ii) Release of radioactivity to the environment.

The fault tree analysis has shown that loss of lives of personnel resulting from accidental exposure to radiation is insignificant.

For the events leading to the release of radioactivity to the environment, fault tree analysis has shown that the probability values for such events occurring are in the order of 4×10^{-2} to 1×10^{-2} per year for drilled emplacement and the penetrator method respectively.

When the number of canisters involved in each event are considered, the drilled emplacement process would result in the loss or misplacement of 2.7 canisters per year on average, while the figure for the penetrator process is 0.2 canisters per year.

The comparisons between the two methods of disposal shows that the penetrator method is in general a safer operation, ignoring any post emplacement differences. The addition of redundant lowering systems to the drilled emplacement process could have the effect of halving the failure rates given above.

ACKNOWLEDGEMENT

Contributions have been made to the technical contents of this report by the following members of Taylor Woodrow staff:

Mr. I. Barody
Mr. M. R. C. Bury
Dr. C. Fleischer

Contributions from outside consultants are acknowledged in Section 7.3.

1. INTRODUCTION

Taylor Woodrow Construction Ltd. completed a feasibility study for the offshore disposal of radioactive waste by drilled emplacement method in March 1984. This work was carried out under CEC Research Contract No. 256-81-WAS-UK, and DOE Contract No. PECD/7/9/034, and is described in Reference 1.

In parallel with this study, Ove Arup and Partners carried out a feasibility study for an alternative method using penetrators. This work was under CEC Contract No. 394-83-7-WAS-UK and DOE Contract No. PECD/7/9/063 - 75/82, as is described in Reference 2.

The study that is subject to this report concerns an investigation into the reliability of both of the above methods of disposal. This work has been performed under contract in the framework of the European Atomic Energy Communities cost sharing programme on radioactive waste management and disposal, CEC Research Contract No. 395-83-7-WAS-UK, and DOE Contract No. 7/9/168 - 133/83.

This report gives a summary of the assessment of both methods of disposal. The work is limited to the reliability assessment of the disposal operation between the stages of dockside handling and the final emplacement in deep ocean sediments. Key assumptions under which this study has been carried out are outlined in Sections 5 and 10.

Sections 4 to 8 cover the reliability study related to the drilled emplacement method and in section 9 the reliability of the penetrator method is discussed.

2. OBJECTIVES

The reliability study has been carried out with the following objectives :-

a) To define major hazards and events on which the assessment would be based.

b) To develop a suitable methodology for assessing the reliability of each method of disposal.

c) To identify crucial stages of the operations and key components which play a major role in the success or failure of the operation.

d) To investigate the sensitivity of the reliability values to changes in various key parameters.

e) To identify means of improving the reliability of the systems.

f) To provide a set of data which would enable a systematic comparison between the various disposal options.

g) To discuss the reliability targets (acceptance criteria) and to compare them with the values which have been computed.

h) Make recommendations for future work and studies.

3. METHOD OF APPROACH

In this section, the method of approach for achieving the stated objectives is outlined.

For a realistic and meaningful assessment of each operation it has been essential to realise that all proposed disposal methods are at the conceptual stage and any technique adopted for their assessment should be compatible with the developed state of the systems at present time.

Two basic methods have been adopted for the assessment of each operation. These are:-

a) Failure Mode Effects and Criticality Analysis (FMECA).
b) Fault Tree Analysis (FTA).

Both methods are described in detail in Sections 6 and 7 respectively. The combination of the two techniques has helped to carry out both qualitative and quantitative assessments of the operations. For the penetrator option, it has been sufficient to carry out only a fault tree analysis.

The assessment of each option has been based on investigating their reliability with respect to two major undesired events. These events are referred to as the "top events" and are:-

a) Release of Radioactivity of the Environment.
b) Loss of lives, related to the Operation Personnel.

The reliability study is focussed on investigating ways in which these events may occur.

The quantitative assessment results in working out a probability value, or rate of occurrence, for these events. An attempt is also made to include a measure of "severity" for each case. This approach would enable a better assessment and comparison of various incidents, which would result in one of the two specified top events.

The investigation of the "probability" and "severity" values related to the two major undesired events leads to the need for establishing a criteria for comparison with other hazards and judgement on their acceptability. Establishing a criteria for acceptance is outside the scope of this study. However, because of the importance of this factor in assessment of waste disposal operations, this subject is discussed in Section 12 and reference is made to a method by which all reliability values related to operations of this nature may be assessed.

4. SUMMARY OF OPERATION

4.1 Description of the Drilled Emplacement Method

Disposal of high level waste by drilled emplacement method involves lowering canisters containing waste into holes drilled in deep ocean sediments and plugging the upper section of the holes.

High level waste, in vitrified form, is contained in canisters and is delivered in fully shielded containers, termed "flasks" in this report, to a purpose built port facility. The flasks are transported by a special vessel to the disposal location where they are transferred to the floating platform.

A semisubmersible platform, probably a concrete caisson type, is stationed at the disposal site to receive the flasks and to carry out the lowering and emplacement of canisters into the predrilled holes.

The handling and assembly of canisters into a string of pipes is carried out inside a shielded and flooded compartment of the platform, known as the "wet cell." The assembly of pipes containing the canisters is lowered by pipes into the holes. Five trips are needed to complete the emplacement for each hole.

A rope system has been considered as an option for introducing redundancy into the lowering system and improving its reliability.

For the convenience of assessment, the operation is broken down into the following stages :-

1. Dockside handling of flasks at port.
2. Transfer of flasks to transport vessel.
3. Transport of flasks at sea, from port to platform.
4. Transfer of flasks from the supply vessel to the platform.
5. Handling of flasks on board the platform and transfer to wet cell.
6. Handling canisters in wet cell.
7. Assembly and forming a string of canisters for lowering.
8. Lowering the string of canisters by pipes.
9. Remote entry of canisters into the hole.
10. Lowering and emplacement of canisters inside the hole.
11. Removal of the top section of casing and plugging the hole. See figure 1.

The operation is expected to be carried out in suitable weather windows with stringent control and monitoring during all stages, in accordance with an approved procedure.

Full details of the operation are given in the final report by Taylor Woodrow (Ref. 1).

4.2 General Parameters

The general parameters related to the drilled emplacement method for the purposes of this study are assumed to be as follows:-

- Total quantity of waste, per annum	400 m^3 (max).
- Total number of canisters, per annum	2700
- Total number of canisters per flask	19
- Total number of flasks per voyage (9 trips per annum)	16 (304 canisters)
- Weight of each flask, full	150 te
- Number of canisters per hole	300
- Number of canisters per lowering trip	60
- Total number of holes per annum (Type 1 waste).	9

General dimensions of type 1 waste canisters :-

- Canister outside diameter	430 mm
- Canister length	1300 mm
- Glass weight	0.36 te
- Canister weight (full)	0.45 te
- Approximate inside dia. of each cased hole	630 mm
- Approximate dia. of uncased hole	915 mm
- Total depth of each hole	1000 m
- Complete duration of emplacement operation per hole (approximate)	24 days
- Total number of personnel involved (4 groups) (excludes land based personnel)	400-500 max.
- Total number of personnel on platform, at each time (2 shifts)	200-25C max.

Disposal Location for Study Purposes

- CV2, Cape Verde Rise
- NAP, Nares Abyssal Plain

See Figure 8.

Effect of location on the overall reliability is discussed in Section 12.

5. REFERENCE CRITERIA - KEY ASSUMPTIONS

The reliability assessment of the disposal operation is based on a number of key assumptions. The effect of some of these assumptions could be significant particularly when various methods of disposal are compared with one another.

The following is the list of major factors and assumptions considered in this study :-

5.1 The method of disposal considered for the reliability study is based on the system described in References 1 and 2. A summary is given in Sections 4 and 9.

5.2 The equipment and components considered for this operation are assumed to be comparable with those developed and used for similar tasks in other industries. The stringent specifications and quality requirements for the equipment used in disposal operations is expected to improve their reliability and performance.

5.3 The systems developed for the disposal operation are expected to be subjected to a comprehensive programme of testing and field trials before they are approved for this purpose.

5.4 The operation is based on handling vitrified waste amounting to a maximum of 400 m^3 per year. Significant changes to the quantity may affect the method of operation to a degree that the reliability values computed for the proposed method would no longer be valid.

5.5 All incidents which result in a temporary halt to the operation or cause any delays are ignored. Such interruptions may increase the chance of some undesired events occurring. Failure rates for most events are taken high enough to take account of this effect.

5.6 It is possible that within the next two decades or so, a major breakthrough in technology may influence the details of this disposal method and the reliability of the entire operation.

Advances which relate to the development of new material or equipment should be considered in any future design or assessment of the disposal operations.

5.7 It is assumed that all health and safety rules imposed by the regulatory authorities, both in the nuclear industry and offshore industry, would be observed.

5.8 No financial restraint has been imposed on the cost of the operation. The system is, however, based on a sensible approach from the point of view of economics and operation logistics which relate to the capability to handle the specified annual rate of waste at reasonable costs.

5.9 A high level of monitoring and control is considered for the operation. The monitoring system is particularly essential for the lowering operation in order to check the behaviour and the state of the lowering pipe system. It has not been possible to include the effect of this factor directly on the reliability assessment. It is anticipated that in the course of any detailed design study and field trials, the effect of the control and monitoring system can be quantified.

5.10 Values which are quoted for the probability or frequency of occurrence of the events are based on available statistics on the performance of similar components used in other industries. Engineering judgement has been used to modify some values for the assessment of this operation.

5.11 The reliability assessment relating to the exposure of high level waste to the environment ignores the effect of introducing an overpack for canisters which may guarantee waste containment for a significant time, e.g. 500 to 1,000 years.

5.12 The study ignores any post-emplacement events which may lead to the release of radioactivity to the environment. The events include earthquakes, tectonic movements and disturbances which result in exposing the canisters to the ocean environment.

5.13 The effect of sabotage on the operation is ignored.

5.14 In the absence of any established method for assessing the "severity" of the consequence of the undesired events, a method has been suggested for including this effect in the overall reliability assessment. This is based on the ratio of the number of canisters lost in an event to the total annual number of canisters involved.

5.15 The effect of disposal site on the reliability of the operation has been ignored. The choice of site may affect the reliability in the following ways :-

a) Environmental conditions.

b) Transport of radionuclides in ocean waters

c) Post emplacement influence of seabed sediments.

5.16 Burial of canisters in depths which are less than those specified are treated the same as in the case of canisters left on the ocean bed. The effect of insufficient burial depth is therefore not taken into account because of lack of sufficient knowledge of its exact consequence.

5.17 The probability values selected for the occurrence of the undesired events are not exact and a closer examination of each case may result in a considerable change to some values.

5.18 The quantitative assessment of the reliability is for the purpose of arriving at an approximate "order of magnitude". Many changes to the probability values are not therefore likely to affect the overall assessment of the operation at this stage of its development.

6. FAILURE MODE, EFFECTS AND CRITICALITY ANALYSIS (FMECA)

6.1 Definition

Failure mode and effects analysis is a technique by which deviation from normal operation or failure modes of components are identified and the consequence of such events are investigated. The basic method is to identify each fault mode and its cause within the system and to trace it to the final consequence(s) or effect(s) on the operation.

The analysis of the system from "cause" to "effect" is often referred to as the "bottom up" technique, as all single failures are addressed first as the main events and are traced to the final consequence.

There are a number of variations to this method and each technique is tailored to suit the system which it is intended to assess.

Criticality Analysis

FMEA techniques may be extended to include a limited assessment of the criticality of each failure mode.

The "criticality" is a measure of combined effect of probability of occurrence of an incident and the severity of the consequence. The severity may be assessed in terms of loss of lives, economic losses, unavailability of the system and exposure of radioactivity to the environment or other hazards.

The inclusion of "criticality" analysis in the FMEA method has resulted in referring to this technique as Failure Mode Effects and Criticality Analysis (FMECA).

6.2 Description of Analysis

For assessment of complex systems with a great number of components, full FMECA becomes extremely tedious. The difficulty is increased even further when the scheme is at its conceptual stage and full details of many components are not available.

Disposal of high level waste by drilled emplacement method is a complex operation and to make the most efficient use of available resources, FMECA has been carried out only for major components and undesired events with the following objectives in mind :-

- To identify the critical incidents which may result in a major loss of lives or exposure of high level waste to the environment.
- To use FMECA tables of failure modes as a checklist for constructing fault trees, discussed in Section 7.
- To use criticality factors in order to identify major critical events and comment on ways of improving the reliability or performance of the related components.

There are basically two methods of approach to FMECA techniques: "functional analysis" and "hardware analysis". In functional analysis, specific tasks and deviations from the performance target are investigated, while in hardware analysis, failure of components is first considered and the consequence is investigated for each case.

Both methods are acceptable and their suitability depends on the type of system and available details for each component. It has been convenient to combine both methods for this study.

6.2.1 Procedure

The following steps have been taken to carry out the analysis:-

- Define each stage of operation and sequence of work.
- Identify and list all major components and their performance requirements.
- Identify major failure modes or deviation from normal operation, which may result in the occurrence of undesired events.
- Trace the "cause(s)" and "consequence(s)" of each failure mode.
- Suggest mitigating factors or corrective measures.
- Evaluate the crticality of each event.

The outline of the above data for each failure mode is presented in tabular form for ease of reference.

The complete set of FMECA tables have been prepared and are issued as an internal report. A selected set of tables from the report is given in Appendix A.

6.2.2 Overall Ranking

Three factors have been selected for criticality ranking. These are :-

F1 Probability of initiating failure.
F2 Probability of specific consequence.
F3 Severity factor.

Since each factor is based on a logarithmic scale, the overall ranking is obtained by adding the three factors.

Factor F1

Component failures are often quoted in terms of failures per 10^6 hours. Since one year is approximately 10^4 hours, the failure rate of 1 in 10^6 hours relate to an occurrence of once in every 100 years or 0.01 per year.

To simplify the ranking procedure, the frequency of each operation within a year is ignored with the exception of emergency actions which occur only "on demand."

The following factors represent (F1) probability ranking:-

F1 FACTOR	ANNUAL RATE OF FAILURE	GENERAL RANKING
5	1 to 10	Frequent
4	0.1 to 1	Probable
3	0.01 to 0.1	Occasional
2	0.001 to 0.01	Remote
1	0.0001 to 0.001	Extremely unlikely
0	0.00001 to 0.0001	Negligible.

It is extremely tedious to select exact values of F1, particularly when the project is at its conceptual stage. The values are therefore approximate and are based on the best engineering judgement and the reliability of similar componnents or operations in other industries.

Factor F2

In some cases, a failure mode represented by F1, does not necessarily result in the occurrence of the event defined as the "consequence." The probability of the specific consequence depends on mitigating factors and simultaneous occurrence of other events.

Factor F2, therefore, represents the probability of the defined "consequence" when the initial failure mode has occurred. The following ranking scale is used to represent the F2 factor:-

FACTOR F2	PROBABILITY OF CONSEQUENCE
5	Consequence inevitable, probability = 1
4	Probability 0.1 (1 in 10)
3	Probability 0.01 (1 in 100)
2	Probability 0.001 (1 in 1000)
1	Probability 0.0001 (1 in 10,000)

Where the effect of a failure is mitigated by another action or emergency system, factor F2 may represent the probability that the protection system fails to operate or to be effective.

Factor F3

Consequences are ranked by reflecting the severity of the event in terms of cost, loss of lives or release of radioactivity to the environment. Release of radioactivity to the environment includes both short term and long term effects relating to unacceptable doses of radionuclides.

F3 Factor	Order of Cost	Examples of Consequence
5	£1000 M	Catastrophic damage to ship or platform - Release of radioactivity - Loss of 100 lives or more.
4	£100 M	Major damage to ship or platform - Release of radioactivity not causing major damage to health - Loss of 10 lives or more.
3	£10 M	Less severe damage resulting in cessation of operations for about a month - Minor release of radioactivity not causing serious danger to health.
2	£1.0 M	Minor damage resulting in cessation of operations for a few days - No release of radioactivity.
1	£100 K	Minor damage resulting in less than one or two days cessation of operation.
0	£10 K	Inconsequential.

The following table is a suggested range for categorising the criticality of consequence. The overall ranking values are the result of adding the three "F" factors.

Overall Ranking Range	Comments/Acceptability
12 - 15	Unacceptable critical value. Action is essential to reduce risk.
10-11	Scope for risk reduction to be considered. Risk may be unacceptable.
1-9	Risk insignificant but factors other than crticality should be examined.

6.3 Summary of Results

Most failure modes have an overall ranking between 5 to 8. A significant number lie in the middle range of 9 to 10, and a few are in the critical range of 11 to 15.

The failure modes which lie in the range of 11 to 15 are :-

Failure mode	Overall Ranking
- Breakage of lowering pipe (various causes)	12
- Supply vessel sinking in open sea	11
- Transfer bridge impact with supply vessel	11
- Undesired unlatching of the remote disconnect joint	11

Breakage of the lowering pipes has the highest overall ranking.

6.4 Assessment of Results

There are a number of ways to introduce an overall ranking value for the undesired events and their consequences. All methods involve combining the ranking values for the probability of occurence and the "severity" of the events. Whatever the method, the values are approximate and have a limited significance.

The criticality analysis of the operation serves the purpose of highlighting the crucial stages of the operation and the critical events and it should be used with the said limitations in mind.

The overall ranking values related to drilled emplacement operations indicate that the three stages of transport at sea, transfer at sea (vessel to platform) and the lowering operation are all crucial and contribute to bringing the overall criticality values to an unacceptable range.

7. FAULT TREE ANALYSIS (F.T.A)

7.1 Definition

Fault tree analysis entails specifying major undesired events which are referred to as the "top event," and constructing a logic diagram which lists all the events which may lead to their occurrence. The "top event" may occur as the result of the combination or simultaneous occurrence of a number of events. Alternatively individual events may cause its occurrence. The construction of the fault tree results in tracing all events down to the primary causes of failures, known as the "basic events." This analytical approach is often known as the "top down" technique as a major undesired event forms the top of the tree and branches lead to the original faults and failures.

There are a number of variations to the F.T.A. method, each suitable for a particular application. "Hardware approach" and "functional approach" are the two common methods. In the hardware approach, failure of individual components is considered, whilst in functional approach, the operation or output of the relevant components is analysed. In circumstances where the system is the combination of established off-the-shelf equipment and newly designed components, the combination of the two techniques is appropriate. This approach is adopted for the disposal operation for ease of analysis.

7.2 Description of FTA Tasks

The following work has been carried out for the reliability study using the FTA technique :-

- Breakdown of operation into distinct stages
- Listing major components used in each stage.
- Examining each stage of the operation and identifying potential failure modes
- Defining the two "top events" of interest to FTA.
- Constructing the fault tree for each top event.
- Estimating the failure rates and/or frequency of occurrence of events (statistical data acquisition).
- Qualitative analysis.
- Quantitative analysis.
- Sensitivity analysis.
- Search for improvements and comments on the results.

Information gathered from the failure mode and effects analysis has helped to identify potential failure modes and sequence of events which lead to the top events.

The top events selected for evaluation by FTA technique are :-

a) Release of radioactivity to the environment.

b) Hazard to personnel involved in the operation.

These two events form the basis of the reliability study for the disposal of high level waste.

7.2.1 Qualitative Analysis

The qualitative assessment involves constructing a fault tree to represent the sequence of events which could lead to the top, undesired event. From this fault tree, a list of groups of basic events is compiled, with each group containing those and only those basic events which if they occurred together would inevitably lead to the top event. These groups are known as the "minimum cut sets".

7.2.2 Quantitative Analysis

The quantitative analysis entails designating probability values or rate of occurrence to each basic event and computing the overall probability for the occurrence of the top event.

The top event may have a different degree of severity depending on the nature of the basic events. The severity for this method of disposal is represented by the number of lives lost or the quantity of high level waste exposed to the environment.

In order to include the severity effect, the probability values for events which reflect the main incidents of loss of lives, or loss of radioactive waste, are multiplied by dimensionless values which represent the proportion of the loss with respect to the total number or quantity involved.

In the case of loss of lives, this number will be represented by the number of lives lost divided by the total number of personnel involved in the operation; and the severity of the high level waste exposed to the environment is represented by the number of canisters lost in each incident, divided by the total annual number of canisters handled.

Inclusion of the "severity" factor helps to represent the reliability value in a more appropriate way and avoids the indiscriminate grouping of all undesired events without taking account of their magnitude or severity.

7.2.3 Sensitivity Analysis

The probability values for the basic events are not exact and some are suspect of being inaccurate. This limitation is inevitable and applies to all cases whose related data may not be accurate or fully reliable for a host of reasons. It is therefore helpful to investigate the effect of each probability value on the overall reliability and have a clear knowledge of the effect of any changes to each value on the overall reliability.

Each top event is represented by a number of minimum cut sets. Each cut set represents one independent chain of events which result in the occurrence of the top event. It has been shown in Appendix C that an "importance value" can be calculated for each basic event which represents the level of its contribution to the overall probability value. The events with high importance value are therefore those which require closer investigation, as changes in their probability factor will have a significant effect on the overall reliability value.

The sensitivity analysis entails computing the importance values for each basic event and working out the effect of predicted changes in the probability factors for these events on the overall reliability value. Table 2 gives the list of the key events and the relevant sensitivity values.

A further application of the sensitivity analysis is in identifying key events which have the most significant effect on the reliability of the systems. This approach is most helpful when a search is made for the best ways to improve the performance and the reliability of the system.

7.3 Sub-Contract Work

Assistance has been sought from outside consultants in two particular areas of work. These are :-

i) Acquisition of statistical data for the fault tree analysis.

ii) Assessment of monitoring and control system related to the lowering and emplacement operation.

7.3.1 Acquisition of Statistical Data

As discussed earlier, the disposal of high level waste involves a number of offshore operations and use of components which are, in many ways similar to some other offshore activities on which statistical data is available. The statistical data relate to the performance of essential components and the frequency of failures and occurrences of undesired events.

The Systems Reliability Service (SRS) of the United Kingdom Atomic Energy Authority were asked to provide statistical data from their data bank and to assist in simplifying the fault tree to suit the type and limitations of the available data. Their computer program ALMONA has also been used to analyse the overall fault tree as a check on Taylor Woodrow's in-house program 'CUTS'.

The statistical data, with relevant explanatory notes and the fault tree for the top event of "release of radioactivity to the environment" are presented in Appendix B (Reference 3).

7.3.2 Assessment of Monitoring and Control System

The success of the disposal operation by the drilled emplacement method relies heavily on continous control and monitoring of the operation during the lowering and emplacement phase of the work. The monitoring relates particularly to the state of the lowering pipes which are the weak link within the system.

Work has been subcontracted to J. McKee & Partners (Services) Ltd. and included:-

- Examining the general monitoring and control system for the lowering operation as recommended by Taylor Woodrow.
- To assess the instrumentation and the back-up system for monitoring and control of the operation.
- To comment on the impact of the various monitoring systems on the overall reliability of the operation.

The instruments relate primarily to the acoustic systems which are used to monitor the location of the hole, the position of the platform and the position and the state of the lowering pipe system. In addition, systems used in the search and recovery operations (in case of loss of the lowering pipe system) are investigated.

Acqusition of data via a cable and by telemetry is also investigated.

A report is issued which summarises the result of this investigation. (Reference 4).

7.4 Assessment of Probability Values

The key factor in the quantitative assessment of the reliability is the selection of correct values for the probability or the rate of occurrence of the "basic events." As discussed in Section 7.3, some data has been provided by SRS of the U.K. Atomic Energy Authority. Most figures relate to shipping accidents and failures connected with the equipment used in the offshore oil industry.

The data of such nature which has been collected over several years relate to components which have undergone major changes and improvements and are likely to be improved even further for this application.

With these limitations in mind, some probability values have been modified to relate them to this particular application. However, in general, because of the presence of so many unknown factors, all probability values used for the fault tree analysis are somewhat conservative. It is clearly understood that some of the selected values may be inaccurate by factors as large as ten or more. The quantitative assessment is therefore made with these limitations in mind, and is used only to obtain an approximate order of magnitude for the reliability of the operation. A sensitivity analysis has therefore been made to investigate the importance of each failure or basic event, and to identify those which changes to their probability value would affect the overall probability most.

7.5 Summary of Results

7.5.1 Overall Probability Value

The summary of the fault tree analysis is given in the following documents :-

- Appendix B. FTA and explanatory notes on derivation of failure rates. Fault tree summary and overall annual failure rates, using computer program ALMONA at Systems Reliability Directorate.

- Table 1 gives the summary of the fault tree analysis, using Taylor Woodrow's computer program CUTS.

- Table 2, the summary of the sensitivity analysis, giving the importance values for each basic event and changes to the overall probability value caused by selected changes to the probability figures of some of the key basic events.

The slight difference in the probability values for "release of radioactive waste to the environment" using ALMONA and CUTS computer programs is because of the numerical rounding off carried out by the ALMONA program.

Two sets of figures are presented for each FTA, one represents the overall annual failure rate without the inclusion of "severity" or quantity of waste involved; the second set introduces the severity by multiplying the relevant failure rate values by the ratio of the number of canisters lost to the total number of canisters handled per year.

7.5.2 Sensitivity Results

The "Importance values" worked out for each event are used to assess the effect of changes in the failure rate of one or more basic events on the overall probability factor.

Using the importance values, the effect of any changes to one or more basic events can be worked out for the top event. This method is explained in Appendix C.

To investigate the effect of some key basic events, changes have been made to their probability values and new overall failure rates for the top event have been worked out. The events of particular interest are :-

- Failure of the lowering pipe system
- The recovery of lost canisters or flasks from the ocean bed
- Inclusion of the redundant rope system
- Transportation of waste at sea - sinking of supply vessel
- Transfer of flasks at sea (vessel to platform)
- Sinking of the platform.

The effect of major changes in the probability values related to these events are investigated and the results are given in Table 2.

The results clearly indicate the importance of the redundant rope system, failure of the lowering pipes and the effect of difficulties associated with the recovery of any lost canisters from deep waters.

It is, however, worth noting that despite the significant effect of changes to some factors, all probability values for the top event are high and are in the range of 0.447×10^{-2} to 4.36×10^{-2} per year. These values are clearly beyond the acceptance limit as discussed in Section 12.

7.6 Effect of Redundancy in the Lowering System

In the feasibility study of the drilled emplacement method (Ref 1) introduction of a rope system as a back-up to the lowering pipe assembly was suggested.

The redundant rope is connected to a point above the string of canisters such that if the lowering pipe breaks during the operation, the canister strings are supported by the rope and are prevented from falling onto the seabed.

Introduction of the redundant rope is likely to cause some complications in the operation and slow down the trips. The full effect of the inclusion of the rope system can not be investigated at this stage of the development, but it is clear that it will be able to have an effect on the undesired events which lead to the loss of canister strings during the lowering operation.

For the purpose of assessment of the redundant rope system, the fault tree has been modified to reduce the probability of the loss of canisters by pipe failure by a factor of 10. This factor implies that because of the presence of the redundant rope, only 1 in 10 cases of the lowering pipe failure would result in the catastrophic loss of canister strings.

The effect of the rope system on the overall reliability is to reduce the probability of release of radioactivity to the environment from 0.987×10^{-2} to 0.447×10^{-2} per year. This is equivalent to a reduction factor of 2.2 for the top event.

7.7 Assessment of FTA Results (Drilled Emplacement Method)

In Table 1 a summary of results, based on a number of assumptions is given. The reliability values range between 0.447 x 10^{-2} to 4.36 x 10^{-2} depending on the use of a redundant rope system and the assumptions associated with the recovery of lost canisters. So far as the reliability is concerned, the overall probability value of 0.987 x 10^{-2}, which allows for some success in the recovery operations from deep water, can be broken down into the following major events :-

PERCENTAGE CONTRIBUTION TO THE OVERALL PROBABILITY

Main Stages of Operation	Case 1 Some recovery considered in deep water	Case 2 No recovery from deep water	Case 3 No recovery, severity included
Losses contributed by lowering operation	84	95	82
Loss contributed by transport at sea (both shallow and deep water)	4	1	5
Loss contributed by transfer at sea	7	3	8
Loss contributed by platform sinking	5	1	5

In all cases, the lowering operation is the main contributor to the overall top event, other events being of significantly less importance.

8. FAULT TREE ANALYSIS HAZARD TO PERSONNEL

8.1 Definition

Hazard to personnel is defined as loss of lives per annum and relates only to the personnel who are likely to be involved in the operation. For the fault tree analysis, "loss of life" is considered as the "top event", and is divided into two categories:-

a) Loss of life caused by exposure to radiation.
b) Loss of life caused by other accidents (non-radioactive).

Fatalities related to the exposure to radiation include all incidents in which either whole body exposure or a concentrated dose would lead to the death of the personnel and excludes the delayed health effects. In Appendix D some basic information is given on the main types of radioactive particles with reference to the doses which may be fatal. This information is not complete and is quoted in this form only as a matter of interest.

Fatalities caused by non-radioactive accidents include those occurring at any stage of the operation, from transport at sea to the final emplacement stage, and includes major hazards such as the sinking of the transport vessel and the platform which would result in a major loss of life.

8.2 Description of Fault Tree Analysis

The fault tree, with the top event of "loss of life" represents all the sequence of events which lead to such a consequence.

In order to present the results compatible with other available statistical data, the probability of occurrence of each incident is multiplied by the severity of fatality ratio which is represented by dividing the expected number of fatalities by the total number of personnel involved in the entire operation, per year.

It is anticipated that some 250 people would be involved at a time on the platform and the supply vessel, and as two groups are assumed to be involved on a rota basis, the total number involved would be in the order of 500 people per annum.

The probability values selected for the occurrence of each event are based on available statistical data and, using engineering judgement, they have been modified to relate them more closely to this particular operation. Figure D-1 in Appendix D represents the related fault tree. All events and their probability values are presented in Appendix D with explanatory notes on their selection. (Data has been drawn from References 5, 6 & 7).

Taylor Woodrow's computer program CUTS has been used to carry out the fault tree analysis. The output is presented in Appendix D.

8.3 Summary of Results

The overall probability rate of fatality is computed to be 0.615

person per year, for a total personnel of 500. A closer look at the major source of contribution to this figure reveals that practically the entire value is contributed by "non-radioactive" related incidents. The probability value for fatalities caused by radioactive related accidents is only 4.2×10^{-4} persons per year or approximately 4 to 5 persons per 10,000 years.

The evaluation of the importance factors reveals the following :-

- Nearly 50% of fatalities are caused by operational activities not involving radiation.
- Approximately 37% are caused by accidents on the personnel transport vessel (excluding sinking of the vessel).
- 8% is due to platform sinking.
- 4% is due to ship sinking.

A full list of importance factors and failure rates is given in Appendix D.

8.4 Assessment of Results

For assessment, the results are best broken into the general categories of radioactive and non-radioactive related incidents.

Fatalities caused by non-radioactive accidents can be compared with the statistical data available from other industries.

Some of the values are presented here for perspective.

Industries	Risk of Death p.a	Risk/Year/Person	Source
U.K. offshore operations (oil and gas industry) 1982-84 combined. Excludes transport and shipping	1 in 2000 approx.	5×10^{-4}	Ref. 6
U.K. offshore operations (fixed and mobile platforms 1982-84 combined. Excludes transport and shipping	1 in 1621	6.16×10^{-4}	Ref. 6
Lives lost due to accidents on board ships and vessels (1975-80) (world-wide)	1 in 2172	4.6×10^{-4}	Ref. 7
Waste disposal operations (fault tree analysis) (non-radioactive related)	1 in 813	1.23×10^{-3}	Appx. D

Waste disposal operations (radio-active related)	1 in 1186000	8.4×10^{-7}	Appx. D

See also Table D-2, Appendix D.

A comparison of these figures indicate that the highest fatality risks related to the disposal operation are primarily the result of the operation being carried out offshore and risks related directly to handling radioactive waste are practically negligible.

The risk values computed by the fault tree analysis are somewhat pessimistic and may be lower by a factor of two or more for the following reasons :-

- The data used for the fatalities on the platform are the same as those related to the offshore oil industry, which includes also hydrocarbon related incidents.

- Because of the nature of the disposal operation, stringent control on the entire operation is likely to reduce the chance of many accidents which often relate to lack of maintenance and care of equipment, and proper procedures on board many vessels and platforms.

- Accidents on board the supply vessel and the passenger vessel and the probability of its sinking, are likely to be less than those used for the fault tree analysis. The FTA values are the same as the worldwide statistical data for fatalities, covering vessels of all types and groups. Available data on ship accidents show that the rate for those registered in Group B (Liberia, Greece, Panama and Cyprus), is much higher than the Group A, traditional maritime states (U.SA, U.K., Japan etc.), both in the absolute number observed and the percentage of total trading ships affected. (Ref. 8).

  For transport of flasks and the personnel, the vessels would be part of the Group A registered ships, with additional safety facilities which should contribute to a safer operation.

The values used for the fault tree analysis can therefore be safely taken as those related to both personnel transport vessel and the vessel which transports the flasks.

It is apparent that most operations become more hazardous if they take place offshore and because of this effect, offshore disposal operations are likely to be more hazardous to personnel, compared with land based disposal operations. It should, however, be pointed out that the risk to personnel as the result of handling radioactive waste canisters and flasks is minimal and is within the permitted level given by a number of regulatory authorities.

Hazard to personnel is not therefore expected to be a key issue in assessment and acceptance criteria for the offshore waste disposal operation of any type.

8.5 Historical Review of Radiation Fatalities

Hazard or risk here in Chapter 8 refers to fatality occuring almost at once but certainly within a month from the time of accident, be it either conventional or critical nuclear radiation. It is important to draw attention to this, especially regarding radiation dose uptake since the somatic, long term and hereditary effects are not considered in this study for any kind of accidents.

C.C. Lushbaugh et al (Ref. 9) give an overview of all serious and fatal radiation accidents worldwide: the accidents, type, effect, location, after treatment and developments are all listed. The only country which is excluded from these statistics, for unknown reasons, is the United Kingdom.

The data collected are those between 1945 and 1979. During this 35 year period, there were 98 radiation accidents involving a total of about 560 persons, of whom 16 died due to radiation effects. Also 3 deaths occurred at Idaho Falls in 1961 due to blast injury in which 3 people from the rescuing team had received and survived Total Body Irradiation (TBI). There are altogether 39 TBI injuries, 29 internal exposure and many more 'local' injuries during this period with various after effects.

The statistics do not state the total number of personnel directly involved with these radiation works.

In order to calculate the risk of fatality, one could assume that the total number of personnel (excluding administration) involved was 10,000. (This includes all countries except the U.K.).

The risk can thus be calculated:-

16 deaths due to irradiation
350,000 total number of personnel assumed

$$r = \frac{16}{350,000} = 4.6\ E\text{-}5\ y^{-1}.$$

or $r = 1$ in $21875\ y^{-1}$.

The radiation accidents were classified according to the type of radiation-producing device or substance involved:

		Number of Accidents	
A.	"CRITICALITIES"		
	a) Critical assemblies	5)	
	b) Reactors	5)	14
	c) Chemical operation	4)	
B.	"RADIATION DEVICES"		
	a) Radioisotopes 'Sources'	41)	
	b) X-ray devices	15)	61
	c) Radar generators	1)	
	d) Accelerators	4)	
C.	"RADIOISOTOPES"		
	a) Transuranics	11)	
	b) Fission Products	4)	
	c) Radium Spills	1)	23
	d) Diagnosis and Therapy	7)	
		Total:	98

From this classification it is evident that accidents occur much more frequently from inadvertent exposure to radioisotopic 'sources' or commercial irradiators, and X-ray generators used in research and analytical work in labortories, than from all other exposure.

9. RELIABILITY ASSESSMENT OF PENETRATOR METHOD

9.1 Summary of Operation

The operation involves loading high level waste canisters into projectiles which are launched at a suitable deep ocean site. The projectiles (penetrators) fall freely through the water and embed themselves completely within the seabed sediments. (Figure 3).

An option exists to include an overpack as part of the penetrator body which would secure containment of the vitrified waste for 500 to 1000 years before the direct exposure to surrounding sediments takes effect.

The penetrators are designed to reach a sufficient penetration depth. The predicted range is between 40 to 70 metres depending on a number of factors. The entry hole is expected to close behind the penetrator, although this phenomenon needs to be verified.

Various scenarios have been proposed for the operation, most of which relate to the stages prior to the launching of the penetrators at canister source and taking the penetrators to the dock, in shielded transport flasks.

The stages of the operation may therefore be summarised as follows (Option E) :-

- Penetrators are assembled at canister source.
- Penetrators are transported by rail in shielded flasks to the port handling facilities.
- Flasks are transferred from the dock to the disposal vessel.
- Flasks are transported to the ocean disposal site by the vessel.
- Each flask is taken by a travelling portal crane to a launching frame.
- The penetrators are removed from the upended flask and are launched through the ship moonpool.

Option E and other options are presented graphically in Figure 2.

9.2 General Parameters

The general parameters related to the penetrator method for the purpose of this study are assumed to be as follows:-

- Total quantity of waste per annum	400 m^3(max).
- Total number of canisters per annum,	2700
- Total number of canisters per penetrator	5
- Total number of penetrators per voyage	18
- Required minimum tail embedment	10 m
- No. of ships	3
- No. of voyages per ship	10

General Dimensions of HLW Canisters

Canister, outside diameter	430 mm
Canister, length	1350 mm
HLW volume	0.15 m^3
Weight, including HLW	0.45 - 0.48 te
Weight, if ullage filled with lead	0.820 te
Pre-disposal land storage	50 years

General Dimensions of Reference Penetrator

Penetrator, outside diameter	650 mm
Penetrator, overall length	8500 mm
Penetrator tube (steel) thickness	75 mm
Penetrator weight (void filled with lead)	18.65 te

Disposal Area for Study Purposes

GME, Great Meteor East
CV2, Cape Verde Rise
SNAP, Southern Nares Abyssal plain
See Figure (8)

10. REFERENCE CRITERIA

The following are the key criteria which form the basis of the reliability assessment :-

10.1 The method of disposal is based on the operation described by Ove Arup and Partners in their report entitled "Ocean Disposal of High Level Radioactive Waste - Penetrator Engineering Study." (Reference 2).

10.2 The equipment used for the operation is based on present capabilities and the state of technology. No outstanding development has been considered which would affect the findings of this study significantly.

10.3 The penetrator method would be employed only if trials showed that the penetrators could reach the minimum depth of penetration necessary to prevent escape of radionuclides. In this reliability study, the failures related to unsufficient depth of penetration are assumed to be caused by some malfunction or departure from the norm, e.g. damage to stabilising fins.

10.4 Variations to the method of operation as specified by Ove Arup and Partners in their report may affect the final severity values which reflect both the quantity of waste involved in an undesired event, and the probability of its occurrence. These variations are not considered in this study, although the sensitivity study related to the fault tree analysis enables the investigation of the effect of most changes.

10.5 The following points which are raised in Section 5 (Reference Criteria - Drilled Emplacement option), also apply to the assessment of this method :- Sections 5.3, 5.7, 5.8, 5.10 to 5.17.

11. FAULT TREE ANALYSIS - PENETRATION METHOD

11.1 General

A general description of the fault tree analysis and the method of approach are given in Section 7. The top events for the reliability study are similar to those for the drilled emplacement operation, and are :-

a) Release of Radioactivity to the Environment

b) Hazard to Personnel.

For the reliability assessment of the penetrator option, only a fault tree analysis has been carried out. It is believed that both quantitative and qualitative assessments, using this method, are comprehensive and are sufficient for comparison of the two methods of offshore disposal, or for any other future assessment.

Hazard to personnel, in comparison with the drilled emplacement method, is less severe. This is because the total number of personnel involved in the operation is less than that for the drilled emplacement method, and the operation is generally of a simpler nature.

The reliability study for hazard to personnel related to the drilled emplacement option has shown that this event is not of great significance for the final assessment of the waste disposal methods, and for this reason, no further work has been carried out on risk to personnel for the penetrator method. The reliability study for this case is therefore limited to studying the risks associated with the release of radioactivity to the environment.

11.2 Fault Tree Analysis - Release of Radioactivity To the Environment

The fault tree, representing the sequence of events which lead to the release of radioactivity to the environment is given in Appendix E.

The fault tree has been kept as simple as possible and the basic events have been limited to the level that meaningful and reliable values can be used for the probability or frequency of their occurrence. The general limitations which have been outlined in Section 7 - Fault Tree Analysis for the Drilled Emplacement Method, also apply to this case and for the purpose of brevity, they are not repeated.

Reliability values for release of radioactivity to the environment are worked out using Taylor Woodrow's computer program CUTS. The summary output is presented in Appendix E.

To enable a direct comparison of reliability values for the various methods of disposal, the quantitative assessment has been extended to include the "magnitude" or "severity" effect. This has been achieved by multiplying the annual failure rates for the key events which lead to loss of high level waste canisters by a factor which is the ratio of the expected number of canisters lost to the total number of canisters handled each year. The results are presented in Appendix E.

11.3 Sensitivity Analysis

A sensitivity analysis has been carried out to investigate the effect of changes in probability values for some key events on the overall reliability value. The summary of the importance values appear in Appendix E.

11.4 Summary of Results

Table 3 gives the summary of the fault tree analysis. The overall probability for the occurrence of the top event, release of radioactivity to the environment, is 1.3×10^{-2} (or 1.3 events in 100 years). The reliability value, with severity or magnitude included, is 0.2 canisters per year (or 20 canisters in 100 years).

As discussed in Section 7, because of uncertainties associated with the recovery of canisters or penetrators when they are lost in deep waters, a pessimistic view of "no recovery" has been taken for the base case.

The importance values, presented in Appendix E, indicate that the event of damage to fins which may result in the unsatisfctory embedment of penetrators, is the main contributor to the overall reliability value, the importance value being 0.76. However, when the severity effect is included, the sinking of the supply vessel in the open seas becomes the key event with an importance value of 0.66. The penetrator imperfections has an importance value of 0.264 in this case.

The contributions to the overall failure rates due to various causes are indicated in the table below.

PERCENTAGE CONTRIBUTION TO THE OVERALL PROBABILITY

Contributing Factor	Case 1 No recovery of penetrators from deep water	Case 2 Some recovery of penetrators from deep water	Case 3 No recovery from deep water severity included
	% of events	% of events	% of canisters
Unsuccessful Embedment	87	89	30
Transport at Sea	13	11	70

Like the drilled emplacement option, faults during the actual disposal operation contribute the greatest proportion of events leading to release of radioactivity, but unlike the drilled option, these events do not lead to the greatest proportion of canisters lost.

12. DECISION ANALYSIS/ACCEPTANCE CRITERIA

12.1 General

To assess the reliability of high level waste disposal operations and for setting conditions which would form the basis for their acceptability, two factors may be considered as the basic criteria. These are :-

a) Performance
b) Safety aspects.

Performance targets relate to the capability of the system to perform all specified tasks. The main factors which relate to performance targets are the cost and the capability of the system to handle the required quantity of high level waste, within a timescale.

The feasibility studies carried out by Taylor Woodrow and Ove Arup and Partners on the two methods of offshore disposal indicate that both options meet the performance targets with some small difference in the overall cost which is not of immediate interest to this study.

Safety aspects are the key factors for projects of this nature and in order to arrive at a decision or pass judgement on the acceptability of any method of disposal, a well defined safety target and criteria are needed. Such criteria would eventually be declared by international regulatory authorities as the basis for acceptance.

Realising that no operation is a hundred percent safe, and that no system is fully reliable, undesired events are bound to occur at some stage of the operation. The key factors which affect the acceptability of such events are :-

- The "magnitude" or "severity" of the consequences
- The probability or expected rate of occurrence.
- Other factors such as justification, social impact, duration of exposure etc., which will be addressed separately.

For the disposal of high level waste in deep ocean sediments, the events which are of prime concern are :-

a) Loss of life related to the personnel involved in the operation.

b) Release of radioactivity to the environment and the consequences.

12.2 Loss of Life

Loss of lives, in this case, relate to those which may occur during the operation and involves basically the personnel who are actively engaged in this work.

The "severity" which is represented by the number of fatalities and the "rate of occurrence" can be combined to give an overall measure of the criticality in terms of fatalities per year, or fatalities per year per number of people involved.

The regulatory authorities may recommend a fatality limit as the peak acceptable value. Such a figure should not be significantly different from that experienced in the offshore oil industry, but, because of the sensitivity of this issue, the acceptable limit is likely to be smaller than many comparable operations which do not involve handling radioactive materials.

For perspective, fatality rates related to a number of industries in the U.K. are given in Table D2 (Appendix D). These rates are all based on available statistical data. (Reference 11).

The fatality rates computed for the disposal operation compare well with those related to many industries. It should, however, be realised that the major portion of the risks associated with the disposal operation results from the fact that the operation takes place offshore and fatalities caused by exposure to radioactive waste are practically negligible.

12.3 Release of Radioactivity to the Environment and the Consequences

The risk associated with the release of radioactivity to the environment should be assessed with the following key issues in mind :-

- Probability or frequency of occurrence
- Severity or measure of harmfulness associated with the consequence (health effects)
- The duration of the release of radioactivity or exposure
- The number of people involved
- Other deleterious effects not associated with health such as causing restrictions in the use of some areas or products.

12.3.1 Probability Limits

Probability or frequency of occurrence has little significance without being related to a measure of severity. In general, higher probability values are accepted for events with low severity consequences and low probability values are expected to be associated with high levels of severity. Acceptance limits associated with various levels of severity are discussed later, but examples of statistical rates related to some catastrophic, non-radioactive incidents are quoted, from Reference 12, in the table below for perspective :-

Accidents	Probability (per yr)
Offshore platforms, catastrophic events resulting in loss of offshore installation, large scale loss of life and/or uncontrollable pollution	$10^{-6} < P < 10^{-4}$ (improbable or low)
Offshore platforms, major events (accident condition) resulting in significant structural damage (repairable), several lives lost, large scale pollution	$10^{-4} < P < 10^{-2}$ (remote or medium)
Offshore Platform Blowout	
Probability per platform per year (N.Sea 1983 Statistics)	3×10^{-4}
Topside fire/explosion leading to major structural damage (N.Sea 1983 Probability per platform per year	1×10^{-4}
Aircraft Industry	
Probability limit of failure for the under-carriage on commercial aircraft, per landing	10^{-7}

The probability limits related to the release of radioactivity to the environment would be linked to the severity or the level of harmfulness of the consequences of such events.

12.3.2 Severity Factors

A number of options exist for quantifying the severity or the magnitude of exposure to radiation. A method used in the nuclear industry relates the severity to the effective dose equivalent which may be taken by an individual in terms of mSv/man/year or other similar units. To arrive at such a figure for accidents related to the disposal operation, a complex model is required to represent the exact conditions which lead to the final dose uptake. Factors which affect the dose uptake the most, are :-

- Location of exposure (ocean floor, Atlantic or Pacific Ocean, etc)
- Habits and diets of individuals.

The effect of location whether deep or shallow water, is primarily dictated by the different environmental features and pathways related to each place. Once the geological barrier is removed or the temporary containment provided by the body of canisters or penetrators ceases to exist wholly or in part, radionuclides will be dispersed directly into the sea as the result of leaching, and the level of radioactivity will rise depending on the process of dispersion.

The transport and mixing of oceanic waters is extremely complex. Figure.7 shows the major phenomena such as internal waves, abyssal circulation, advection, convection, etc. which vary depending on the location, and affect the mixing time and the level of concentration of radionuclides. The complete dispersion which eventually results in a peak concentration of radionuclides may take tens to several thousand of years.

For computation of the maximum dose taken by man or other creatures, mathematical models are needed to represent the effect of all factors including the complex chemical and biotic processes. Creating a model which represents the complete and true phenomenon of dispersion and dose uptake requires full knowledge and verification of each process. In addition, the sensitivity of the overall model to the variations in these factors, needs to be thoroughly investigated. The complexity of the dispersion process makes the forming of an accurate mathematical model somewhat doubtful and because of lack of sufficient knowledge in many related areas, pessimistic values with high safety factors, may be selected for each process. The level of confidence on the accuracy of the models therefore plays a significant role in recommending acceptable limits for the probability and dose uptake values.

12.3.3 Dose Uptake

The dose uptake may be from the following sources :-

- Ingestion of contaminated aquatic biota (fish, seaweeds, etc)

- Ingestion of contaminated seawater and salt
- Inhalation of sea spray.
- Exposure to shore sediments
- Immersion in contaminated water

Through these sources a certain dose may be taken by individuals depending on their dietary and other habits.

The absorbed dose should be "weighted" to represent its level of harmfulness. The weighting factor should represent :

- the level of harmfulness of different types of radiation (X-rays,   particles, etc)
- the degree of harm to different organs and tissues.

The weighted dose is referred to as the "Effective Dose Equivalent" which reflects the above effects and represents the true level of its harmfulness. Conversion factors are available to work out the effective dose equivalent (Ref. 11) for most types of exposure.

12.3.4 Computation of Risk

For computation of risk, the following steps are therefore to be taken :-

1-Calculating the probability of occurence of the top event.

2-Calculating the probability of occurrence of each event which results in the release of radioactivity to the environment. These events are presented as minimum cut sets in the fault tree analysis.

3-Creating models for representing the consequences of the undesired events for the purpose of quantifying the severity.

4-Creating models for people's habits for the purpose of working out accurate maximum dose uptakes and associated probabilities.

5-Modifying dose uptake to obtain the effective dose equivalent which represents their actual level of harmfulness.

The outcome would present a number of scenarios for exposure which would be defined by the probability values and the effective dose equivalent uptake related to the maximum exposure condition.

12.4 Acceptance Criteria

12.4.1 Objectives

The objective for setting up criteria for acceptability of waste disposal or other radioactive related operations is to provide a clear guideline and limit for acceptability of such operations. The acceptability in this case is primarily related to hazard to people (health effects) based on

established radiological protection objectives. Other effects such as restricting the use of an area or products also play a part in acceptance criteria, but their role is less significant.
According to the National Radiological Protection Board (NRPB) (Ref.13), at the present time there are no comprehensive internationally agreed radiological protection criteria which can be applied to all solid waste disposal options.

The NRPB recommends that radiological protection objectives for solid waste disposal should have the following features :-

- They should be based on "risk" where risk is defined as the combination of the probability that particular exposure situations will occur and the probability that subsequent doses will give rise to deleterious health effects.

- There should be a requirement that the risk to an individual should not at any time exceed a specified level.

- There should be a requirement that total risk (i.e. risks to populations) should be as low as reasonably achievable (ALARA), economic and social factors being taken into account (optimisation of protection).

- No practice shall be adopted unless it produces a positive net benefit (justification).

  This requirement relates to an entire practice, such as electrical power generation and its associated nuclear fuel cycle, of which waste disposal is a component.

So far as risk and dose limits are concerned, there are a number of points which should be considered and taken into account whenever a criterion for their acceptance is to be established. These are :-

a) Correct "dose" calculation, covering the effect of various types of radiation, short term and long term exposure, concentrated local body exposure, overall body exposure, time or duration effect and the harmfulness effect, most of which can be presented in terms of "effective dose equivalent."

b) Uncertainty factors and effects. Uncertainty relates to many factors and includes :-

- Uncertainties and inaccuracies associated with models which represent the dispersion media and pathways to man.

- Uncertainties associated with selected probability values for the occurrence of the top event.

- Uncertainties associates with dietary and other habits of individuals which affect the dose uptake.

- Uncertainties associated with the effect of dose uptake (long and short term).

- Sensitivity of relevant factors and their impact on the overall dose or risk values..

- Uncertainties associated with predicting long term future conditions.

c) Consideration of all exposure scenarios. This implies that risks associated with all phases of waste disposal should be considered globally. The disposal management includes all land based operations prior to transport, land transport, sea transport, disposal operations and long term post emplacement conditions. As the result of global consideration, a modified risk and dose limit may be specified for specific stages of the waste management in addition to the overall global effect.

d) Long term duration effects. Incidents which lead to unstoppable long term consequences should be treated as a more severe case, and risk and dose limits should be modified to reflect their long term impact.

e) Consideration of dose uptake by the public at large in comparison with exposure to radiation by a limited number of people involved in the operation (occupational exposure) or other cases such as medical treatment.

f) Consideration of the "consequence" as an accident situation.

g) Consideration of involuntary dose uptake from other sources (land, cosmic, etc).

h) Justification of the operation as described below.

12.4.2 Recommendations of International Commission on Radiological Protection (ICRP)

The recommendations of ICRP apply to the post emplacement phase of high level waste disposal operations, and apply to the sum of exposures from all sources, excluding natural background radiation and medical radiation.

In order to have some idea of the overall risk limits and objectives, the following statements are quoted from the National Radiological Protection Board (NRPB) publication No. NRPB-GS1 (Ref.13) which refers to ICRP recommendations (Ref. 14).

"The limit on effective dose equivalent recommended by ICRP for individual members of the public is 5mSv per year. However, in cases where prolonged exposures are involved, ICRP suggest that the individual effective dose equivalents should be restricted to 1mSv per year. It is therefore appropriate to define a risk limit, corresponding to the 5mSv dose limit, for general application and a risk objective corresponding to the 1mSv dose objective, for application when doses are likely to be incurred over many years.

"In order to derive the risk limit and objective from the corresponding dose levels, it is necessary to determine the probability per unit dose that a deleterious health effect will ensue. ICRP conclude that for radiological protection purposes the probability per unit dose equivalent that an individual will contract a fatal cancer can be taken to be about 10^{-2} Sv^{-1}, and the overall probability that there will be serious genetic effects in any of an individual's descendants is about 0.8×10^{-2} Sv^{-1}.

"For the purpose of deriving the risk limit and objective it is therefore appropriate to use a rounded factor of 2×10^{-1} Sv^{-1} to take account of both fatal effects in the present generation and genetics in subsequent ones.

The above considerations lead to the following recommendations:-

i) The risk to an individual from waste disposal should not exceed 10^{-4} in a year.

ii) For those situations where doses would be incurred over periods exceeding 10 years, the risk to an individual should not exceed 2×10^{-5} in a year.

"In both cases the risk is defined as the overall probability that a deleterious radiation effect (i.e. a fatal stochastic or non-stochastic somatic effect, or a serious genetic effect) will occur. For perspective, it may be noted that the average risk to individuals in the U.K. from natural background radiation is about 4×10^{-5} in a year."

NRPB also state that the risk limits and objectives apply over the whole period from the end of disposal operations (i.e. from the time at which the repository is backfilled and sealed).

As stated earlier, the above figures are quoted to show an order of magnitude for dose limits and objectives and do not directly apply to the stage of the disposal operation which are of interest to this study.

For information and for appreciating the significance of various dose uptake values, Figure 10 from UKAEA Publication "The Effects and Control of Radiation" (Ref.15) is included in this report. Dose limits stated in NRPB Publication GS1 (Ref.13) and ASP7 (Ref.16) are also presented in figure 10.

12.4.3 Subjective Judgement

The criteria discussed so far, have been basically unprejudiced objectives which are an aid to, and part of, a more complex decision process.

The decision criteria for acceptability of any waste disposal operation would therefore be based on the following factors :-

- Dose limits and dose objectives.
- "Justification" as described in 12.4.1.
- Consideration of long term future impact.
- Consideration of effect on public at large.
- Consideration of other effects, not related to health effects, such as restrictions in use of an area or products.
- Consideration of effect on other creatures.
- Consideration of uncertainties related to present and future processes.
- Consideration of exposure to radiation from other sources.
- Social and political impacts.

The acceptability of the disposal operation becomes therefore a "judgement" task in which the various criteria discussed here play a role but are not on their own sufficient for automatic acceptance of any disposal method.

12.5 Criteria for Comparison of Disposal Methods

For comparison of disposal methods, all factors which are common to the options can be ignored. Consideration of features which clearly differ for the various disposal methods simplifies the process of assessment. Such comparison on its own is not, however, sufficient for acceptance, particularly if the comparison does not cover all stages of the disposal management and operation including the post emplacement effects.

Factors which play a major role in comparison of various disposal methods may be listed as follows :-

- Severity of the consequences represented by maximum effective dose equivalent.

- Probability values related to various "dose uptake" scenarios.

- Cost factors.

- Hazard to operators.

- Effects other than "health effects" such as restrictions in use of an area.

- Post emplacement differences.

13. COMPARISON OF DRILLED EMPLACEMENT AND PENETRATOR METHODS

The numerical comparison between the two methods, for the probability of release of radioactivity to the environment, is summarised as follows:-

Description	Drilled Emplacement Method	Penetrator Method
Overall reliability value (no recovery from deep water) Accidents/year	4.36×10^{-2}	1.31×10^{-2}
Overall reliability value (some recovery from deep water) Accidents/year	0.987×10^{-2}	1.22×10^{-2}
Overall reliability value (with magnitude included) Canisters/year	2.75	0.2

These figures indicate that, so far as the disposal operation alone is concerned, the penetrator method is marginally a safer operation. It should, however be noted that in both cases the probability values are relatively high and, in the light of approximations made at this stage of their development, the difference between the two options is not significant.

The penetrator method is also a safer operation when hazards to personnel are considered. The reason for this difference is that, in general, smaller numbers of people are involved in the operation and that no disposal platform is needed for the penetrator option.

The comparison of the two options indicates that so far as incidents related to sinking of the supply vessel is concerned the difference between the two options is insignificant. However, if events related to the loss of canister strings during the lowering operation and loss of individual penetrators are compared, because of the significant difference between the number of canisters in each case (60 against 5), the maximum likely dose uptake by individuals will be higher by a factor of 12, for the drilled emplacement method. The factor of 12 is approximate and is based on comparing two identical dispersion models for the ocean and radionuclide pathways to man.

Comparison of the two methods for post emplacement conditions is outside the scope of this study. This case should, however, be considered whenever an overall assessment of the disposal management is made or various options are compared on the basis of their entire phases of operation.

14. CONCLUSIONS

In this report, attention has been focused on assessing the probabilities of occurence of two specific undesired events:

a) Hazard to personnel engaged in disposal operations
b) Release of radioactivity to the environment

The estimates of failure rates and probabilities used to derive the overall failure rates are by no means certain. Some values are based on actual reliability data, though not for the specific items of equipment to be used in the offshore disposal system, and some are based on engineering judgement because there is no satisfactory data available. It is important to realise that the numerical values calculated by the fault tree analysis are to be used for guidance only and should not be taken as absolute values.

14.1 Hazard to personnel

The analysis has shown that the radiation hazard to personnel engaged in the disposal operation is insignificant in relation to the normal hazards of being offshore e.g. shipping accidents, falls, drownings etc. These risks are assessed at about 0.6 fatalities per year compared to the radiation hazard of 4×10^{-4} fatalities per year.

14.2 Events leading to some release of radiation

The probability of an event leading to release of radioactivity has been assessed as 1.3 events in 100 years for the penetrator system and 4.4 events in 100 years for the drilled emplacement. In this calculation, no account was taken of the quantity of waste lost, and any canisters misplaced outside coastal waters were deemed to be irrecoverable.

14.3 Quantity of waste misplaced

Assessment of the consequences of waste lost are beyond the scope of this investigation, but an assessment has been made of the number of canisters lost in each event. The calculations indicate that over a period of 100 years, the penetrator system may misplace 20 canisters, while the drilled emplacement system may misplace 275 canisters, in each case out of a total of 270000 canisters.

14.4 Contributory factors to canister loss

The loss of canisters in the drilled emplacement system was almost entirely due to accidents occurring while lowering the waste into the hole. In the penetrator system, the majority of losses were attributable to the ship sinking, with misplacements caused by fin misalignment accounting for about 30% of losses.

14.5 Effect of redundancy

The drilled emplacement system described in Ref. 1 considered the use of a redundant rope to overcome the problems of fracture of the lowering pipe. The use of such a system with an effectiveness of 90% would reduce the number of events leading to release of radioactivity to 1.7 in 100 years, and the number of canisters lost or misplaced to 129. These represent reductions of 62% and 53% respectively.

14.6 Overpacks

It has been shown that if containment of high level waste can be secured for at least 500 years by some kind of an overpack, the overall level of activity of the waste is likely to drop to below 0.1% of its original level after this period. Such a significant reduction would mean that canisters lost in the ocean with such overpacks would limit the level of maximum effective dose equivalent uptake to individuals to below ICRP's level of "insignificance" (0.5×10^{-5} Sv/yr) for all events which result in the loss of canisters in the ocean.

15. LIST OF REFERENCES

1. The Offshore Disposal of Radioactive Waste by Drilled Emplacement: A Feasibility Study. By M.R.C. Bury of Taylor Woodrow Construction Ltd. Published in 1985 by Graham & Trotman Ltd. for the Commission of the European Communities. EUR 9754 (1985).

2. Ocean Disposal of High Level Radioactive Waste - Penetrator Engineering Study. By Ove Arup & Partners, June 1985. Published as Dept. of Environment Report DOE/RW/85.085.

3. The Reliability Assessment of a Proosed Method for the Disposal of Radioactive Waste. Systems Reliability Service U.K.A.E.A. April 1985. (Contractor's Report, not for publication).

4. Reliability of Sub-Seabed Disposal of High Level Waste - Drilled Emplacement Method. Assessment of Monitoring and Control System for Lowering and Emplacement Operation. By J. McKee & Partners (Services) Ltd. May 1985. (Contractor's Report - not for publication).

5. Mortality Rates among Windscale and Calder Workers. E. A. Clough, G.B.Schofield and F.A.Ward. International Symposium on Biological Implications of Radionuclides Released from Nuclear Industries. Vienna 1979. IAEA.

6. Development of Oil and Gas Resources of the UK, 1985. An Annual Report to Parliament by the Department of Trade. H.M.S.O. Publication.

7. Casualties to Vessel and Accidents to Men, 1985. Department of Trade and Transport Joint Yearly Report. H.M.S.O. Publication.

8. Preliminary Pre-Emplacement Safety Analysis of the Sub-Seabed Disposal of High Level Nuclear Waste. Sandia National Laboratory - SAND 83-7105, January 1985.

9. Total Body Irradiation: A Historical Review and Followup Paper by: C.C. Lushbaugh, Shirley A. Fry, Karl F. Hubner, and Robert C. Ricks, Presented at the Proceedings of the REAC/TS International Conference, Oak Ridge, Tennessee, USA, October 18-20 1979, and Published by Elsevier/North-Holland, Inc. in 1980 entitled: "The Medical Basis for Radiation Accident Preparedness".

10. Ocean Disposal of Heat Generating Radioactive Waste. Penetrator Trajectory Modelling and Emplacement Criteria. By Ove Arup & Partners, February 1985. (To be published as DOE report).

11. Living with Radiation. NRPB. 1984.

12. Risk Assessment. Offshore Oil and Gas Operations. By Floyd R. Tuler. Mechanical Engineering, Nov. 1984.

13. Radiological Protection Objectives for the Disposal of Solid Radioactive Wastes. National Radiological Protection Board. NPRB - GS1. 1983.

14. Recommendations of the International Committee on Radiological Protection. ICRP Publication No. 26. Annuals of the ICRP Vol.1, No. 3, 1977. Pergamon Press.

LIST OF REFERENCES (Cont'd)

15. The Effects and Control of Radiation. P.A.H. Saunders. Nuclear Environment Branch, Environmental and Medical Sciences Division, A.E.R.A. Harwell. U.K.A.E.A. July 1981.

16. Small Radiation Doses to Members of the Public. National Radiological Protection Board - ASP 7.

TABLE 1: SUMMARY OF FAULT TREE ANALYSES - DRILLED EMPLACEMENT METHOD
TOP EVENT: Release of Radioactivity to Environment

CASE REF.	DESCRIPTION	OVERALL PROBABILITY VALUE	UNIT	COMMENTS
1 APPENDIX .C..	Base Case - no redundant rope	0.987×10^{-2}	No. of events per annum	In this case some success in recovery of lost waste in deep water is assumed.
2 APPENDIX .C..	Same as base case (1) with no recovery of any waste lost in deep water	4.36×10^{-2}	No. of events per annum	No redundant rope is included in recovery from deep water
3 APPENDIX .C..	As case (2) but with severity effect included	2.75	Porportion of canisters lost	In this case the quantity of waste lost in each relevant incident is included.
4 APPENDIX .C..	A redundant rope is included	0.447×10^{-2}	No. of events per annum	Some deep water recovery is assumed as in case (1)

TABLE 2: SENSITIVITY ANALYSIS - DRILLED EMPLACEMENT METHOD
TOP EVENT: RELEASE OF RADIOACTIVITY TO ENVIRONMENT

Case Reference	Description	Events Affected (see fault tree)	Combined Importance	Overall Failure Rate
A	Base Case - some recovery from deep water			0.987×10^{-2}
B	Failure rate of lowering pipe doubled	A52, A54	1.811	1.787×10^{-2}
C	No recovery waste lost in deep water. Probability of recovery = 0.001	A12, A16, A37 A40, A42, A69	4.417	4.36×10^{-2}
D	Inclusion of redundant rope system. Probability of pipe failure reduced by factor of 10	A52	0.452	0.447×10^{-2}
E	No recovery from deep water and inclusion of redundant rope section	Those in C & D	1.686	1.66×10^{-2}
F	Transport at sea, sinking rate for the vessel halved	A13, A19	0.98	0.967×10^{-2}
G	Stranding of platform failure rate doubled	A85, A86, A87, A91	1.003	0.987×10^{-2}
H	Transfer at sea, accident rate doubled	A15, A22	1.0679	1.054×10^{-2}
I	Sinking of Platform - rate doubled	A68	1.0003	0.987×10^{-2}

TABLE 3 SUMMARY OF FAULT TREE ANALYSIS - PENETRATOR METHOD
TOP EVENT: RELEASE OF RADIOACTIVITY TO ENVIRONMENT

DESCRIPTION	OVERALL PROBABILITY	UNIT	COMMENTS
Base case (Probability only)	1.21×10^{-2}	No. of Accidents Per Annum	No recovery from deep water is assumed
Case 2 (Probability only)	1.14×10^{-2}	No. of Accidents Per Annum	Same recovery from deep water
Case 3. Magnitude on "Severity" Effect Case	0.2	Canisters/year	No recovery from deep water is assumed

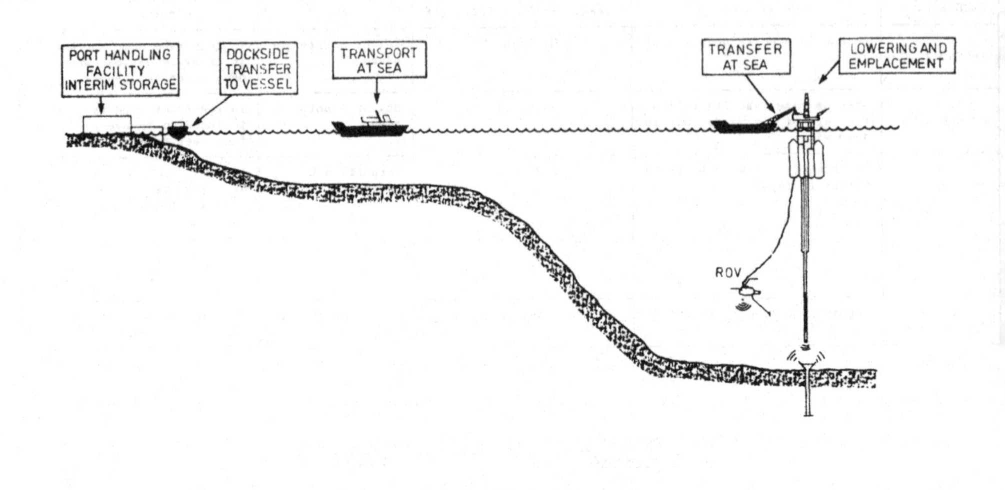

FIG.1. DISPOSAL OPERATION DRILLED EMPLACEMENT METHOD

	SOURCE	transfer	DOCK	load	SHIP	voyage	DISPOSAL SITE
Option A	Ⓒ—Ⓞ—Ⓟ		Ⓢ				Ⓡ
Option B	Ⓒ—Ⓞ		Ⓢ—Ⓟ				Ⓡ
Option C	Ⓒ—Ⓞ		Ⓢ		Ⓟ		Ⓡ
Option D	Ⓒ		Ⓢ—Ⓞ—Ⓟ				Ⓡ
Option E	Ⓒ		Ⓢ—Ⓞ		Ⓟ		Ⓡ
Option F	Ⓒ		Ⓢ		Ⓞ—Ⓟ		Ⓡ

KEY
- Ⓒ Canister storage
- Ⓞ Overpack assembly
- Ⓟ Penetrator assembly
- Ⓢ Dockside storage
- Ⓡ Release

FIG 2. DISPOSAL OPERATION OPTIONS PENETRATOR METHOD

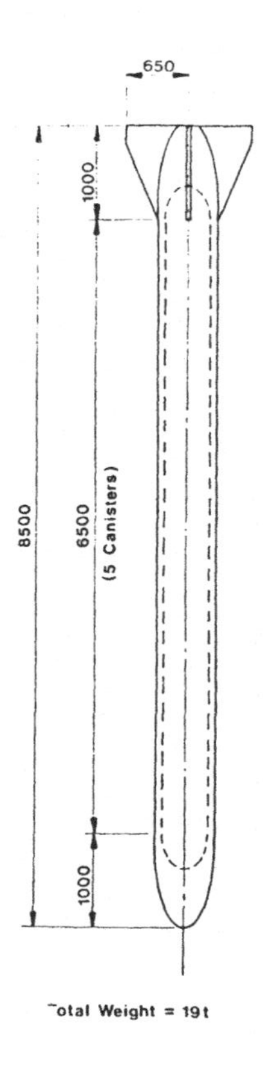

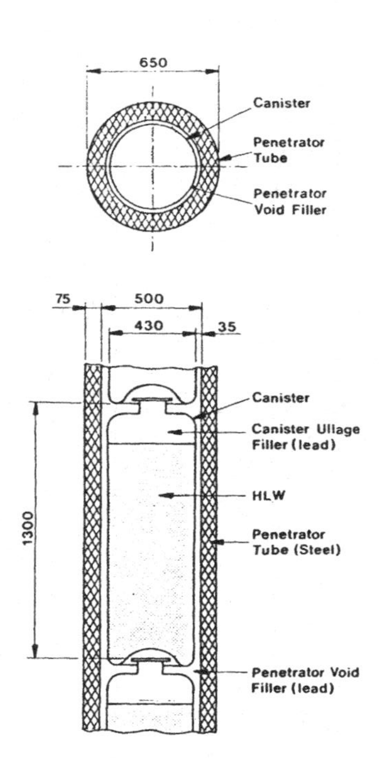

FIG 3. PENETRATOR REFERENCE DESIGN

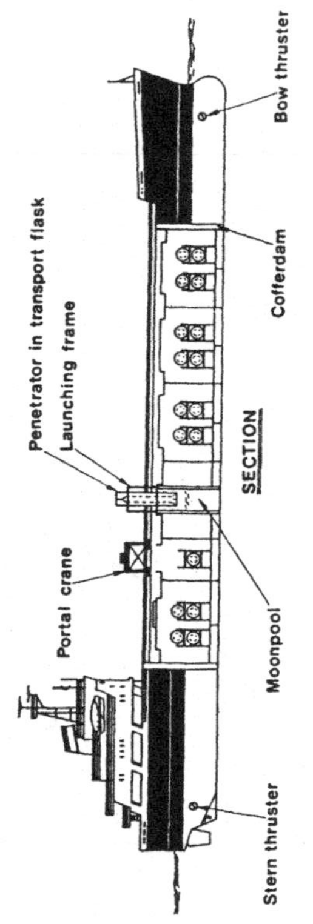

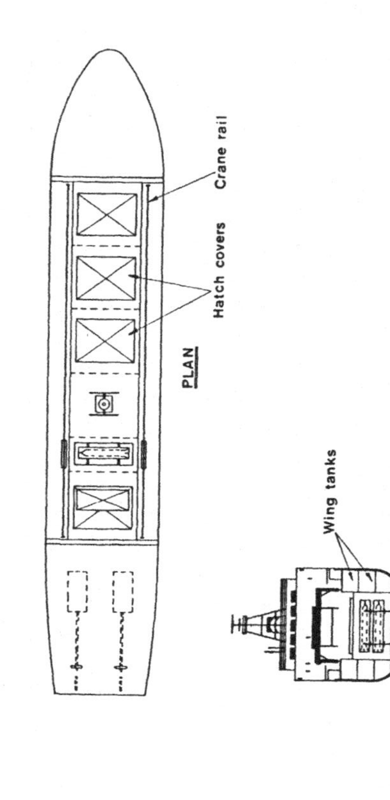

FIG. 4. DISPOSAL SHIP- OPTION E

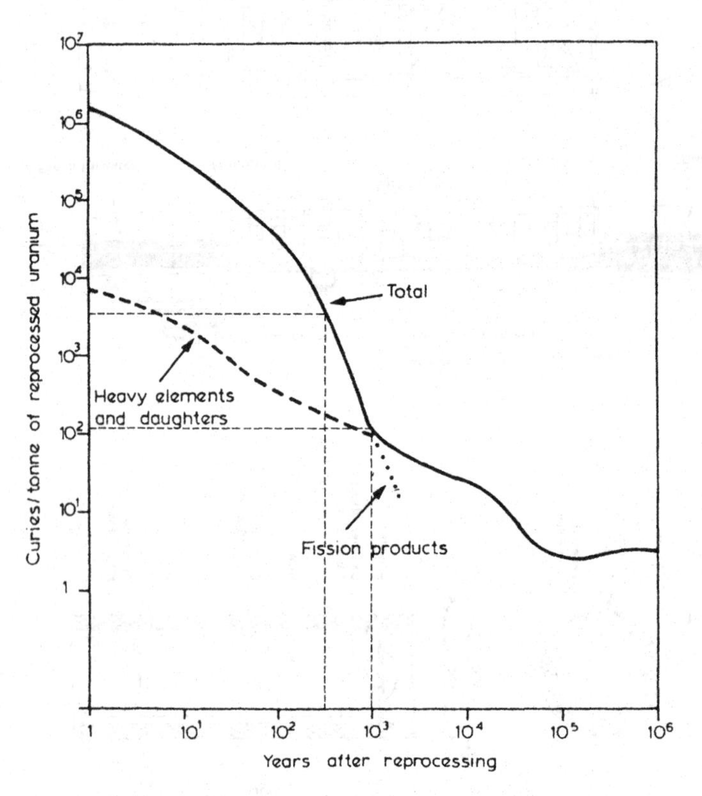

FIG 5. GRAPH OF DECAY OF TOTAL RADIOACTIVITY PER TONNE OF REPRODUCED URANIUM OVER 10^6 (SEARLE,1979)

OPERATION	LOCATION OF RELEASE	ACCIDENT DESCRIPTION	PROBABILITY OF RELEASE IN A 25 YEAR OPERATIONAL PERIOD	EFFECTIVE MAX. INDIVIDUAL DOSE* (Sv/yr.)
Transportation to Port in HLW Barge	River	High-speed collision of a ship into the barge	2.5×10^{-6}	1.3×10^{-5}
Storage in Sealed Cask at Port	Air	Canister dropped in hot cell, particulates released	4.3×10^{-2}	3.0×10^{-10}
Transportation to Disposal Site in HLW Ship	Coastal Waters, Surface	Collision, ship floats, 2 wk release from 6 damaged canisters	1.2×10^{-6}	8.8×10^{-9}
		Grounding, 1 yr release from 6 damaged canisters	1.6×10^{-4}	4.3×10^{-6}
		Collision and grounding, 1 yr release from 12 damaged canisters	2.1×10^{-6}	8.6×10^{-6}
	Coastal Waters, Bottom	Collision, ship sinks, release from 6 damaged canisters, undamaged containers are recovered	3.3×10^{-5}	3.2×10^{-4}
		Collision, ship sinks, no recovery, canisters fail after 100 years	6.0×10^{-6}	7.0×10^{-3}
		Collision, ship sinks, no recovery, release from 6 damaged canisters, remaining canisters fail after 100 yrs	3.3×10^{-7}	7.0×10^{-3}
	Open Ocean, Surface	Collision, ship floats, 4 wk release from 6 damaged canisters	$5 3 \times 10^{-5}$	3.5×10^{-8}
	Open Ocean, Bottom	Collision, ship sinks, no recovery, canisters fail after 100 years	2.5×10^{-4}	Pac: 7.8×10^{-6} Atl: 2.3×10^{-5}
		Collision, ship sinks, no recovery, canisters fail after 100 years	1.4×10^{-5}	Pac: 7.8×10^{-6} Atl: 2.3×10^{-5}

* Dose in the event of an accident (not expected dose).

Fig. 6. Summary of Results of Preliminary Pre-emplacement Safety Analysis (source: Ref 8).

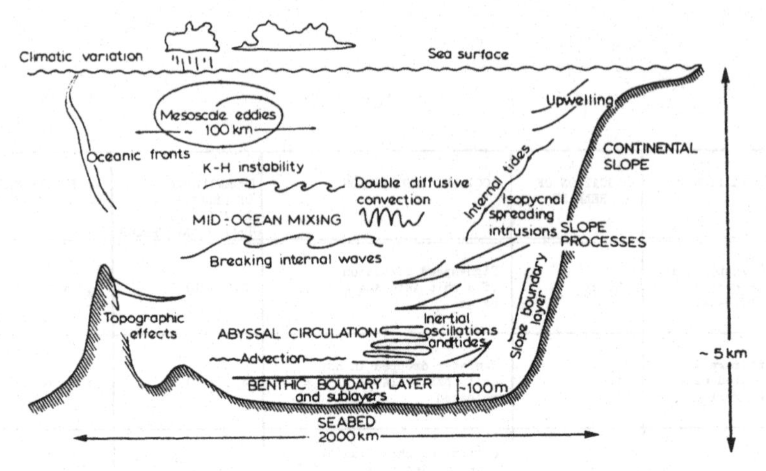

(Source : Ref.10)

FIG.7. PROCESS OF DISPERSION OF RADIONUCLIDES IN OCEAN

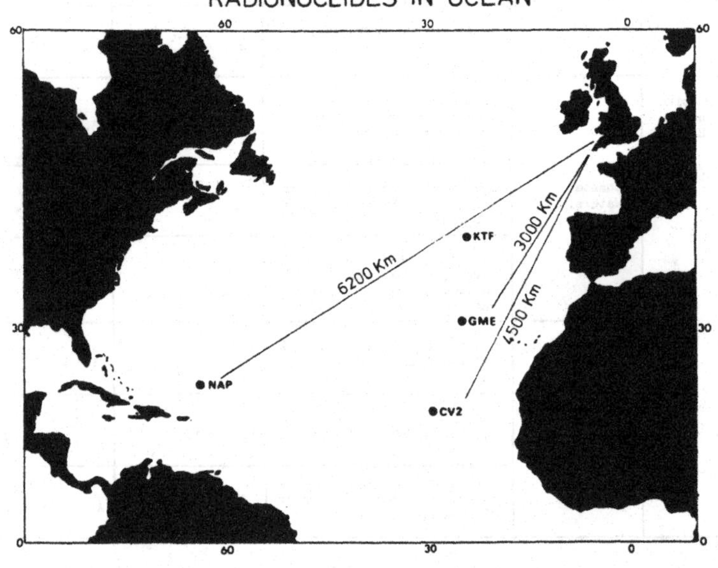

CV2	Cape Verde Rise
GME	Great Meteor Seamount East
KTF	Kings Trough Flank
NAP	Nares Abyssal Plain

From Seventh International NEA/Seabed Working Group Meeting
La Jolla California March 15-19, 1982
SAND82 - 0460 (1982)

FIG.8. POSSIBLE SEABED DISPOSAL STUDY SITES

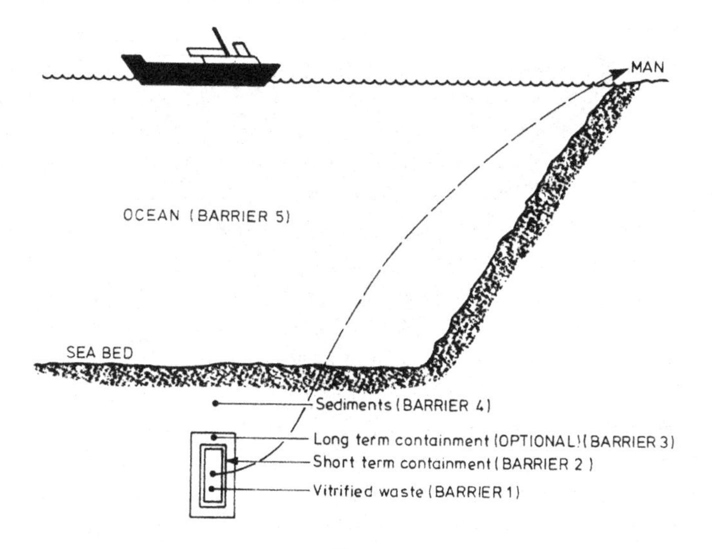

FIG. 9. OCEAN DISPOSAL-
BARRIERS TO MAN

FIG.10

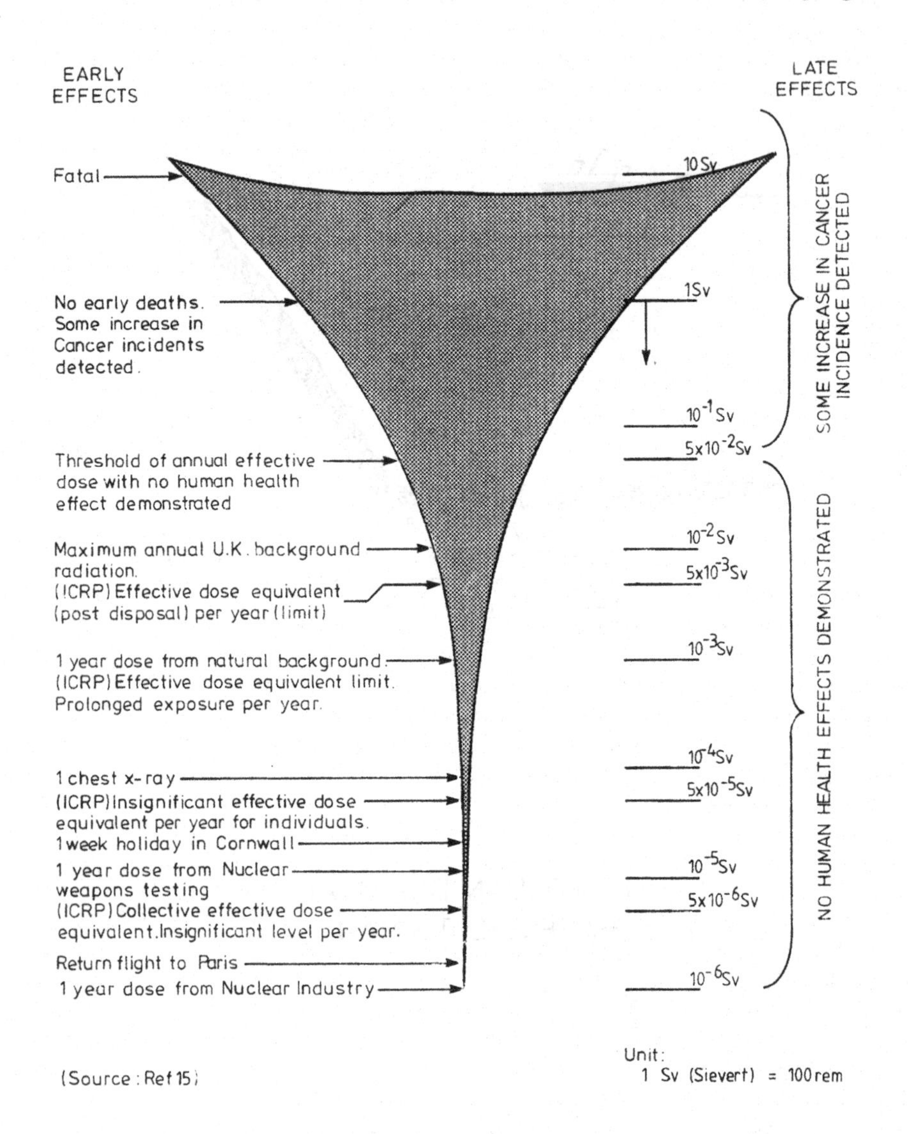

FIG.10. EFFECTS OF RADIATION

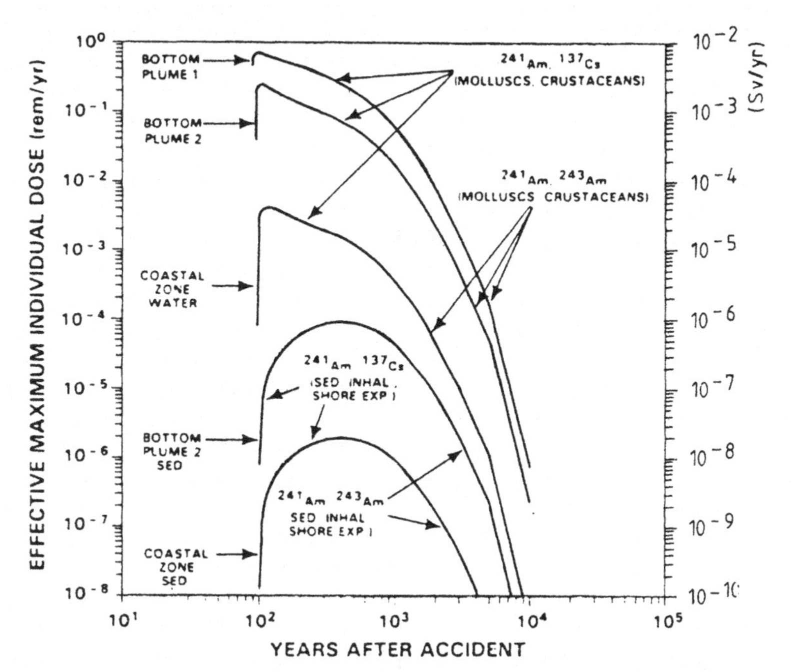

FIGURE 11A. Individual Dose vs Time for Collision and Sinking Accidents in the Deep Pacific Ocean.

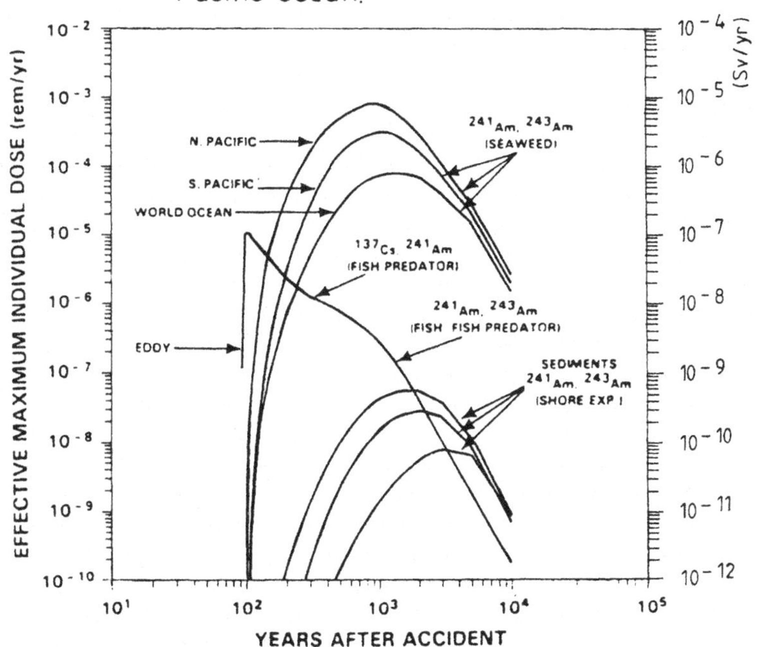

FIGURE 11B. Individual Dose vs Time for Collision and Sinking Accidents in the Deep Pacific Ocean.

APPENDIX A

EXTRACTS FROM

FAILURE MODE EFFECTS AND CRITICALITY ANALYSIS
FMECA

DRILLED EMPLACEMENT METHOD

The example presented here has been selected as an example from Taylor Woodrow internal report "Failure Mode Effects and Criticality Analysis for the Drilled Emplacement Method."

APPENDIX A

SUMMARY OF THE OPERATION: LOWERING AND EMPLACEMENT

CODE: G

1. DESCRIPTION

The lowering and emplacement involves lowering the pipe strings which contain the canisters into the hole. This is achieved by introducing drill pipes which are also used for grouting operations or circulating mud or water into the hole. When the load reaches its downhole destination, the lowering pipe assembly is disconnected remotely from the load and the drill pipes are retrieved back to the rig floor.

2. ASSUMPTIONS

2.1 A semi-automatic pipe handling system is used to handle and assemble the drill pipes.

2.2 The lowering operation (tripping in) is carried out using a pipe jacking system.

2.3 The main hoisting assembly on the derrick is used as the redundant lowering system. The system is also used for remote entry operations when the motion compensation system on the derrick is to be used.

2.4 An option involving the deployment of a polyester/Kevlar rope, as the redundant lowering system, is considered. (See alternative method GA).

2.5 Holes are pre-grouted, using delayed setting grout and, as the result, the difficult task of grouting the casing/canister pipe annulus is alleviated.

3. SEQUENCE OF OPERATION

G.1 In this phase of work the special joint with remote disconnect section is attached to the canister pipe assembly (the load).

G.2 The load is lowered by introducing pipe strings, using the derrick and the pipe handling system on the deck (tripping in).

If the hydraulic jack is used, the hoisting system on the derrick will function as the redundant member and will follow the movements of the jack.

G.3 The remotely operated vehicle (ROV) is launched from the platform and is used to monitor the position of the suspended pipe assembly and to carry out other inspection tasks.

G.4 When the load reaches near the wellhed, with the help of the sonic system and the TVs attached to the ROV, the pipe assembly is positioned accurately above the guide cone. This is achieved by deploying the dynamic positioning system of the platform. The pipe assembly is then lowered gradually into the hole.

G.5 Tripping-in operation continues until the canister pipes reach their destination inside the hole.

G.6 The remote disconnect joint is activated and the load is separated from the suspended lowering pipe assembly.

G.7 If the pre-grouting is not planned to cover all the trips into the hole, then additional pre-grouting operations are carried out for the subsequent rips. The lowering pipe is used to pump the grout into the hole as described in pre-grouting operation.

G.8 Tripping out operation commences by lifting the pipe assembly, unspinning and removing discrete lengths of the drill pipe, using the derrick's hoisting and pipe handling system.

4. LIST OF MAJOR COMPONENTS FOR LOWERING AND EMPLACEMENT

The equipment listed for the hole inspection and pre-grouting operation is also used for this operation. In addition, two other major pieces of equipment may be deployed in order to improve the reliability of the operation. These are:-

a) A hydraulic pipe jack, similar to the system used in the wet cell, but located below the rig floor.

b) The redundant rope system which comprises:

- The polyester/Kevlar paralay braided rope
- Storage bin for the rope
- Traction winch system with tension control safety device
- Pipe/rope connection units
- Power units
- Monitoring and control unit for the system.

DISPOSAL METHOD : Deep Seabed Disposal in Predrilled Holes

OPERATION : LOWERING AND EMPLACEMENT

SUB OPERATION : Making connection with the load

CODE : G.1

Failure Mode		Possible Causes	Potential Consequences		Mitigating Factors	Ranking			
No	Description		Primary	Secondary		F_1	F_2	F_3	O/A
G1-10	PIPE HANDLING SYSTEM Does not start	- equipment failure - malfunction of control system - operator's error	Delay in operation	-	- R.I. & M. - improve Q.A. and control at design and construction stage	5	4	0	9
G1-20	Undesired pipe movement (above rig floor)	- control malfunction -equipment/component malfunction - operator's error	- damage to pipe - damage to equpt. - damage to rig structure -injury/loss of life	-	- R.I. & M. - improve Q.A. and control at design and construction stage. - include failsafe features - include movement monitoring devices - improve equipment layout and protect vulnerable parts	5 5 5 5	5 2 1 1	0 0 0 2	10 7 6 8
G1-30	Pipe drops (above rig floor)	-control malfunction - equipment fault/ failure - operator's error	- damage to pipe - damage to equipt. - damage to rig structure -injury/loss of life	-	As above	4 4 4 4	5 2 1 2	0 0 0 2	9 6 5 8

DISPOSAL METHOD Deep Seabed Disposal in Predrilled holes

OPERATION LOWERING AND EMPLACEMENT

SUB OPERATION Making connection with the load

CODE:G.1

Failure Mode		Possible Causes	Potential Consequences		Mitigating Factors	Ranking			
No	Description		Primary	Secondary		F_1	F_2	F_3	O/A
G1-40	Pipe stops in one position (above rig floor)	-control malfunction -equipt.fault/failure - operator's error	-delay in operation	-	-R.I. & M. - improve Q.A. and control at design and construction stage. - include failsafe features. - include movement monitoring devices - improve equipment layout and protect vulnerable parts. - include manual over-ride.	5	5	0	10
G1-50	Pipe is over spun	- Operator's error -control malfunction -equipment (power tong malfunction)	- damage to pipe joint	-	- R.I. & M. - torque control/ preset limit - torque limit safety features	4	5	0	9
G1-60	Power tong/Iron roughneck does not function	-control malfunction -equipt. malfunction	- delay in operation	-	- R.I. & M. - provide a redundant system	5	4	0	9
G1-70	Remote disconnect joint does not engage	-mechanical failure - operator's error	- delay in operation	-	- R.I. & M. - introduce failsafe feature in the joint - employ experienced operators	5	4	0	9

2

DISPOSAL METHOD : Deep Seabed Disposal in Predrilled Holes

OPERATION : LOWERING AND EMPLACEMENT

SUB OPERATION : Tripping-in (hydraulic jack deployed)

CODE : G.2

Failure Mode		Possible Causes	Potential Consequences		Mitigating Factors	Ranking			
No	Description		Primary	Secondary		F_1	F_2	F_3	O/A
G2-10	Hydraulic jack does not move	-control malfunction - operator's error - equipment malfunction/failure	- delay in operation	-	- R.I. & M. - use experienced operators	5	4	0	9
G2-20	Hydraulic jack is lowered too fast	-control malfunction -operator's error -equipment malfunction/failure -rupture of hydraulic lines	- Damage to hydraulic jack - damage to pipe assembly - overloading of pipe assembly	-delay in operation	- R.I. & M. - use experienced operators -include failsafe mechanism in case of hydraulic pipe rupture -include lowering speed monitoring device -include emergency stopping & damping device	3 3 3	4 3 3	0 0 0	7 6 6
G2-30	Power slips do not close (spider unit)	-control malfunction - operator's error - equipt. malfunction/failure	-entire load may be carried by derricks main hoisting system - delay in operation	-	- R.I. & M. - include manual over-ride - use derrick's main spider as a temporary support	5 5	4 5	0 0	9 10

DISPOSAL METHOD : Deep Seabed Disposal in Predrilled Holes

OPERATION : LOWERING AND EMPLACEMENT

SUB OPERATION : Tripping-in (hydraulic jack deployed)

CODE : G.2

Failure Mode		Possible Causes	Potential Consequences		Mitigating Factors	Ranking			
No	Description		Primary	Secondary		F_1	F_2	F_3	O/A
G2-40	Power slips do not open	-control malfunction -operator's error - equipt. mal-function/failure	Delay in operation		-provide manual over-ride -use derrick's hoisting system to disengage slips - R.I. & M. - use experienced operators	5	4	0	9
G2-50	Drawworks does not function	-control malfunction -equipt. malfunction - operator's error	Major delay in operation	Load transfer to the main hoisting system (overloading likely	- R.I. & M. - include failsafe device on the torque limiter - provide overload alarm system - use experienced operators - provide manual override on the drawworks	5	3	0	8
G2-60	Derrick collapses * (Related to G2, G4 and G5)	-structural weakness - high winds - high motions of the platform	-*Pipe handling and hoisting equipt. collapse.		-R.I. & M. -Use high factor of safety - do not operate when adverse weather conditions are forecast	2	5	3	10
			- Damage to rig floor			2	5	1	8
			-Damage to other equpt. around derrick area			2	5	1	8
			- Lowering pipe breaks (above rig floor)			2	4	2	8
			- Loss of lowering pipe			2	4	3	9
			- Loss of load			2	4	3	9

4

DISPOSAL METHOD : Deep Seabed Disposal in Predrilled Holes

OPERATION : LOWERING AND EMPLACEMENT

SUB OPERATION : Tripping-in (hydraulic jack deployed)

CODE : G.2

Failure Mode		Possible Causes	Potential Consequences		Mitigating Factors	Ranking			
No	Description		Primary	Secondary		F_1	F_2	F_3	O/A
G2-70	Lowering pipe breaks	- fatigue (pipe material) - overstressing of pipes (high bending and deflection) - material fault - joint failure (material fault) - joint unscrews - pipe overstressing (pipe assembly is locked in vibration) - pipe overstressing caused by dynamic loading as the result of the malfunction of the lowering system	- loss of load plus - loss of lowering pipes (load and pipe assembly drop onto the seabed)		- R.I. & M. - limit service life of the drill pipes -use monitoring devices to check pipe deflection and vibration -use emergency procedures to reduce risk or alleviate it (see notes on emergency operation) -avoid operating in adverse environ-mental conditions -use high factor of safety for drill pipes -use superior quality drill pipe -improve Q.A. and control -use a redundant lowering system (rope)	3	5	4	12

5

DISPOSAL METHOD : Deep Seabed Disposal in Predrilled Holes

OPERATION : LOWERING AND EMPLACEMENT
SUB OPERATION : Tripping-in (hydraulic jack deployed)

CODE : G.2

Failure Mode		Possible Causes	Potential Consequences		Mitigating Factors	Ranking			
No	Description		Primary	Secondary		F_1	F_2	F_3	O/A
G2-80	Undesired unlatching of the remote disconnect joint	-control malfunction -operator's error -material failure -faulty equipment	Load drops onto the seabed	-	-Use high factor of safety against failure by over-loading -use failsafe mechanism against control failure/ malfunction -use unlatching mechanism which involves major change or reversal of load on the joint -include more than one change or action to disconnect the joint (e.g. vertical and rotational movement or vertical move-ment plus pressure change) -use a redundant lowering system (rope)	2	5	4	11

DISPOSAL METHOD : Deep Seabed Disposal in Predrilled Holes

OPERATION : LOWERING AND EMPLACEMENT

SUB OPERATION : Tripping-in (Hydraulic jack deployed)

CODE : G.2

Failure Mode		Possible Causes	Potential Consequences		Mitigating Factors	Ranking			
No	Description		Primary	Secondary		F_1	F_2	F_3	O/A
G2-90	Canister pipe breaks	-material fault/ failure of pipe (normal load)	Part of the load drops onto the seabed	-	-use high factor of safety against failure from over-stressing	2	5	3	10
	Also applies to G5 case but load drops into the well	-pipe failure (overstressing)	Load drops into the well (G5)		-use relatively short lengths of load in each trip	2	5	2	9
					-improve material quality assurance and control				
G2-100	Canister pipe joint breaks	-material fault/ failure			-use high factor of safety against failure				
		-joint overstressing	Part of the load drops onto the seabed	-	-improve joint detail to reduce stress condition, etc.	2	5	3	10
	Also applies to G5 case		Part of the load drops into the well (G5)		-improve quality assurance and control	2	5	2	9

7

DISPOSAL METHOD : Deed Seabed Disposal in Predrilled Holes

OPERATION : LOWERING AND EMPLACEMENT

SUB OPERATION : Deployment of R.O.V.

CODE : G.3

Failure Mode		Possible Causes	Potential Consequences		Mitigating Factors	Ranking			
No	Description		Primary	Secondary		F_1	F_2	F_3	O/A
G3-10	R.O.V. launching system does not work	-control malfunction -operator's error -equipt.malfunction	Delay in operation	-	- R.I. & M. - include redundant launching system	4	4	0	8
G3-20	R.O.V. launching system out of control (during launching)	-control malfunction -operator's error -equipt. failure/ malfunction	-damage to R.O.V. -damage to equipt. on R.O.V. -structural damage to launching structure and platform - loss of R.O.V.	Delay in operation	-R.I.& M. -failsafe mechanism included -reduce exposure of equipment - use experienced personnel -keep redundant ROV and equipment	4 4 4 4	5 4 4 2	0 0 0 2	9 8 8 8
G3-30	ROV Equipt., TV camera/sonar trans ponders etc. do not function	-control malfunction -equipt.malfunction - operator's error	- loss of some of monitoring and remote entry aids (operation continues) -delay in operation (replacement of faulty component)	Delay in operation	- use redundant monitoring components - have other independent aids, (e.g. sonar pingers at tip of pipe). - R.I.& M.	5 5	4 4	0 0	9 9
G3-40	ROV's guidance/ maneouvring system malfunction	- control failure - component failure	- minor damage to ROV and equipt. - delay in operation (delay in remote entry	Delay in operation	- use redundant monitoring components - etc. as G3-30	4 4	4 5	1 0	9 9

8

DISPOSAL METHOD : Deep seabed Disposal in Predrilled Holes

OPERATION : LOWERING AND EMPLACEMENT

SUB OPERATION : Remote Entry

CODE : G.4

Failure Mode		Possible Causes	Potential Consequences		Mitigating Factors	Ranking			
No	Description		Primary	Secondary		F_1	F_2	F_3	O/A
G4-10	Dynamic Positioning system malfunctions	-control/guidance malfunction - D.P.power failure - thruster failure	Delay in operation		-have redundant guidance system to check and verify -include direct control over ride feature -have reserve power and thrusters	4	5	0	9
G4-20	Monitoring equipt. TV camera, sonic devices, etc, malfunction or not working	- power failure - physical damage/ loss of equipt. - malfunction of components	Delay	If entry proceeds under this condition: -damage to dummy pipe -damage to monitoring equipt. -damage to guidecone /wellhead	-R.I. & M. -have redundant power supply - include separate monitoring/guidance system (e.g.pingers at tip of dummy pipe).	5 5 5	3 3 2	0 0 1	8 8 8
G4-30	Motion Compensator is locked - not working	-control malfunction - component malfunction - power failure	- delay - damage to dummy pipe - damage to well-head/guidecone - lowering pipe overloaded - lowering pipe breaks (only if entry is attempted -canister pipe drops (only if entry is attempted)	-	-R.I.& M. - support the pipe assembly and lower by hydraulic jack	3 3 3 3 3 3	5 3 3 2 1 1	0 0 0 0 1 3	8 6 6 5 5 7

9

DISPOSAL METHOD : Deep Seabed Disposal in Predrilled Holes

OPERATION : LOWERING AND EMPLACEMENT

SUB OPERATION : Remote Entry

CODE : G.4

Failure Mode		Possible Causes	Potential Consequences		Mitigating Factors	Ranking			
No	Description		Primary	Secondary		F_1	F_2	F_3	O/A
G4-40	Motion Compensator piston is dropped to its lowest position suddenly	-control malfunction -component failure	-piston or support assembly breaks in sudden impact -entire lowering assembly is subject to a dynamic load (NO piston break)	- damage to monitoring equipt -damage to guidecone -lowering pipe overloaded -lowering pipe drops -damage to dummy pipe	- R.I. & M. -fluid emergency shutdown valve inc. -improve strength and structure of vulnerable components	1 1 1 1 1	3 3 3 1 3	1 1 0 2 0	5 5 4 4 4
G4-50	Drawworks operates accidentally	-control malfunction - operator's error - component malfunction	-damage to guidecone -damage to dummy pipe -damage to subsea monitoring equipt. - overloading of lowering pipe -breakage of lowering pipe (loss of load)		- R.I. & M. - use skilled/ experienced operators - include failsafe locking system on drawworks	3 3 3 3 3	4 4 4 5 2	1 0 0 0 4	8 7 7 8 9
G4-60	Pipe lowered outside the hole	- sudden loss of control or mal-function of dynamic positioning system - sudden internal wave or current change acting on pipes - monitoring equipt. error (misreading etc).	-damage to guidecone -damage to dummy pipe -damage to subsea monitoring equipt. -overloading of lowering pipe -breakage of lowering pipe (loss of load)	Delay	-have more than one redundant monitor-ing system -improve the flexibility of dummy pipe -reduce speed of entry -entry under environmentally calm and stable conditions	3 3 3 3 3	4 5 4 4 2	1 0 0 0 4	9 8 7 7 9

DISPOSAL METHOD : Deep Seabed Disposal in Predrilled Holes

OPERATION : LOWERING AND EMPLACEMENT

SUB OPERATION : Lowering into the Hole (Load inside the hole partially or wholly)

CODE : G.5

Failure Mode		Possible Causes	Potential Consequences		Mitigating Factors	Ranking			
No	Description		Primary	Secondary		F_1	F_2	F_3	O/A
G5-10	Lowering pipe breaks	- overstressing caused by:- - excessive drifting of the platform - high subsea currents/waves - material fault or fatigue	Pipe drops partly on seabed and partly into the hole	- hole is blocked or damaged	-continuously monitor pipe deflection	3	4	2	9
				- delay in operation		3	5	0	8
				- wellhead is damaged	-include pipe deflection measuring devices along the pipe.	3	4	1	8
				- load is lost in the hole (fishing req.)		3	5	3	11
				- load is lost on the seabed (in parts)	-resort to emergency operation in case of excessive drifting	3	3	4	10
				- lowering pipe assy. is lost	-use high factor of safety against failure by yielding - R.I. & M.	3	5	2	10
G5-20	Lowering pipe joint breaks	- overstressing of the joint caused by: - excessive drifting - high subsea currents/waves - material fault or fatigue - damaged threads	Pipe drops partly on seabed and partly into the hole	- hole is blocked or damaged	- make joints stronger than pipe	2	4	2	8
				- delay in operation	- continuously monitor pipe deflection	2	5	0	7
				- wellhead is damaged		2	4	1	7
				- load is lost on the seabed (in parts	- including pipe deflection measuring devices along the pipe	2	3	3	8
				- lowering pipe assy. is lost		2	5	2	9
				- load is lost in the hole	-resort to emergency operation in case of excessive drifting. - R.I. & M.	2	5	3	10

11

DISPOSAL METHOD : Deep Seabed Disposal in Predrilled Holes

OPERATION : LOWERING AND EMPLACEMENT

SUB OPERATION : Lowering into the Hole (Load inside the hole partially or wholly)

CODE G.5

Failure Mode		Possible Causes	Potential Consequences		Mitigating Factors	Ranking			
No	Description		Primary	Secondary		F_1	F_2	F_3	O/A
G5-30	Pipe unscrews	Rotation of pipe by various causes	- load/pipe drops on seabed - load/pipe drops (in part) into the well	-hole is blocked or damaged	-include a locking device in the Joint	2	4	2	8
				-delay in operation	-avoid rotating the pipe in the un-screwing direction	2	5	0	7
				-wellhead is damaged		2	4	1	7
				-load is lost in the hole (fishing req.)	- use sufficient number of threads to avoid accidental detachment by one or two turns.	2	5	2	9
				-load is lost on the seabed (in parts)		2	3	4	9
				-lowering pipe assy is lost		2	5	2	9
G5-40	Pipe meets obstruction in the hole	- solids or equipt. left in the hole - casing collapse or damage . grout has hardened earlier than expected	Delay	- need for reaming and surveying the hole	-Pre-inspection by the drilling rig -install soft plug after hole completion to pro-tect hole. - include reaming tool as part of the dummy pipe - washout hole before pregrouting operation	4	4	0	8

DISPOSAL METHOD : Deep Seabed Disposal in Predrilled Holes

OPERATION : LOWERING AND EMPLACEMENT

SUB OPERATION : Remote Disconnection

CODE G.6.

Failure Mode		Possible Causes	Potential Consequences		Mitigating Factors	Ranking			
No	Description		Primary	Secondary		F_1	F_2	F_3	O/A
G6-10	Remote disconnect joint does not operate	-control malfunction -jamming of parts - wrong disconnect procedure	-delay in operation	- need to deploy a second disconnect mode (e.g. controlled explosive)	This joint requires a special design which should fulfil the following requirements: -have no risk of disconnecting by error or mal-function of equipt. or control system. -include at least two actions to achieve dis-connection (a safety requirement) -provide a secondary method of dis-connection if the first method fails -provide a final additional method of disconnection if all normal methods fail (e.g use of controlled charge).	4	5	0	9

DISPOSAL METHOD : Deep Seabed Disposal in Predrilled Holes

OPERATION : LOWERING AND EMPLACEMENT

SUB OPERATION : Grouting

CODE : G.7

Failure Mode		Possible Causes	Potential Consequences		Mitigating Factors	Ranking			
No	Description		Primary	Secondary		F_1	F_2	F_3	O/A
G7-10	Pumping equipment does not start (Mud pumps)	-control malfunction -equipment malfunction	Delay in operation		-have at least two redundant pumps - R.I. & M.	3	2	0	5
G7-20	Pumping equipmnet stops operating (Mud pumps)	-control malfunction - equipment malfunction	Delay in operation		- have at least two redundant pumps - R.I. & M.	3	2	0	5
G7-30	Pipe line bursts or leaks	- material failure - joint failure	Delay		- have by-pass and redundant line - R.I. & M.	3	2	0	5
G7-40	Circulation impossible	- pumping line blockage - blockage within drill pipe	Delay (tripping out and inspection may be necessary)		-clean and test line before operation starts - inspect lowering pipes before assembling - apply higher pressure temporarily to clear line	3	2	0	5
G7-50	Grout injection pump not working	-control malfunction -equipt. failure	Delay		-have a minimum of three pumps - R.I. & M.	3	2	0	5

DISPOSAL METHOD : Deep Seabed Disposal in Predrilled Holes

OPERATION : LOWERING AND EMPLACEMENT

SUB OPERATION : Grouting

CODE : G.7

Failure Mode		Possible Causes	Potential Consequences		Mitigating Factors	Ranking			
No	**Description**		**Primary**	**Secondary**		F_1	F_2	F_3	O/A
G7-60	Surface pipeline ruptures	- Joint failure - damage to pipe - excessive pressure - material fault	Delay		-use spare line - R.I. & M.	3	2	0	5
G7-70	Drill pipe ruptures	- material fault/ failure - overstressing of the line	- loss of drill pipe -need to trip out and change faulty pipe		- use high factor of safety against failure by pressure	2 2	5 2	2 0	9 4
G7-80	Pumps stop working in mid operation	-control malfunction -operator's error - equipt. failure/ malfunction	-delay/reduction in pumping rate -circulation and hole washout. Repeat grouting		- R.I. & M. - have a minimum of three pumps	3 3	2 3	0 0	5 6
G7-90	Grout plugs the drill pipe	-poor/wrong grout mixture -other objects plugging pipe -long stoppage in mid-operation	- pipe ruptures → - tripping out necessary	Loss of pipe Delay in operation	- have high F.O.S. against high pressure bursting - test and control grout	3 3	3 5	2 0	8 8
G7-100	Grout is lost down hole	-poor downhole plugging -casing failure	-delay in operation -need for surveying and logging the hole -need to inject grout ~~to plug the faulty~~ section		-inspect and test log the hole at the end of drilling & casing operation -quality control of casing and cementing operation	2 2 2	5 3 3	0 0 0	7 5 5

DISPOSAL METHOD : Deep Seabed Disposal in Predrilled Holes

OPERATION : LOWERING AND EMPLACEMENT

SUB OPERATION : Grouting

CODE G.7

Failure Mode		Possible Causes	Potential Consequences		Mitigating Factors	Ranking			
No	Description		Primary	Secondary		F_1	F_2	F_3	O/A
G7-110	Plugs do not engage properly	-obstruction to plug seat -faulty plug	-delay in operation -need to trip out and change plug		-design simple, reliable plug and seating arrangements	4 4	5 3	0 0	9 7

DISPOSAL METHOD : Deep Seabed Disposal in Predrilled Holes

OPERATION : LOWERING AND EMPLACEMENT

SUB OPERATION : Tripping Out

CODE : G.8

Failure Mode		Possible Causes	Potential Consequences		Mitigating Factors	Ranking			
No	Description		Primary	Secondary		F_1	F_2	F_3	O/A
G8-10	Breakage of Elevator block Note: This and the following incidents are related to tripping out only. During lowering, the pipe jack system carries the load of the pipes and canisters.	- material/component failure - rupture of links/ pins	-pipe assy drops into the hole (in part) -pipe assy drops onto the ocean bed (wholly or in part)	-damage to rig floor /structure -damage to spider assembly -damage to other equipt. -loss of lowering pipe - Injury/loss of life	- R.I. & M. -high Q.A. and control -high factor of safety -improve design of weak parts	3 3 3 3 3	4 4 3 3 1	1 0 0 2 2	8 7 6 8 6
G8-20	Undesired unlatching of Elevator block	- operator's error -control malfunction	-pipe assy drops into the hole (in part) -pipe assy drops onto the ocean bed (wholly or in part)	-damage to rig floor /structure -damage to spider assembly -damage to other equipt. -loss of lowering pipe in the hole -loss of lowering pipe on the seabed	-R. I. & M. -include load lock safety feature - use skilled operators	IGNORE AS LOAD LOCK SAFETY FEATURE IS INCLUDED			
G8-30	Undesired locking of Elevator block	-control malfunction - operator's error	Major delay in operation		- R.I. & M. - have manual over-ride.	5	2	0	7

DISPOSAL METHOD : Deep Seabed Disposal in Predrilled Holes

OPERATION : LOWERING AND EMPLACEMENT

SUB OPERATION : Tripping Out

CODE : G.8

Failure Mode		Possible Causes	Potential Consequences		Mitigating Factors	Ranking			
No	Description		Primary	Secondary		F_1	F_2	F_3	O/A
G8-40	Various secondary stoppages related to the operation of major or minor components	-control malfunction -component malfunction	Halt to operation	Minor delays	-R.I. & M. - Use skilled operators	5	1	0	6
G8-50	Undesired unlocking of hook	-operator's error -control malfunction	- pipe assembly drops into the hole (in part) - pipe assembly drops onto the ocean floor	-damage to rig floor -damage to spider -damage to other equipt. along path -loss of lowering pipe -injury/loss of life	- R.I. & M. - have failsafe control system - have load sensitive locking device - use skilled operators	IGNORE AS LOAD LOCK SAFETY FEATURE IS INCLUDED			
G8-60	Breakage of hook	-overloading -material fault/ fatigue	- pipe assembly drops into the hole (in part) - pipe assembly drops onto the ocean floor	-damage to rig floor -damage to spider -damage to other equipt.along path -loss of lowering pipe -injury/loss of life	-R.I. & M. -high quality assurance & control - limit service life	3 3 3 3 3	4 4 3 3 1	1 0 1 2 2	8 7 7 8 6
G8-70	Collapse of Crown block	-overloading -material fatigue	Pipe assembly drops	-damage to rig floor and moonpool -damage to spider/ elevator units -damage to other equt. -loss of lowering pipe -injury/loss of life	- R.I. & M. -high Q.A. and control -limit service life	2 2 2 2 2	4 4 3 3 1	1 0 1 2 2	7 6 6 7 5

DISPOSAL METHOD : Deep Seabed Disposal in Predrilled Holes

OPERATION : LOWERING AND EMPLACEMENT

SUB OPERATION : Tripping Out

CODE : G.8

Failure Mode		Possible Causes	Potential Consequences		Mitigating Factors	Ranking			
No	Description		Primary	Secondary		F_1	F_2	F_3	O/A
G8-80	Crown block dislodged or develops a fault	- overloading - material/component failure	delay in operation	-minor damage to equipt. on rig floor	- R.I. & M. - limit service life	2	3	1	6
				-minor damage to drill pipe		2	3	1	6
G8-90	Elevator/Winch cable breaks	- material fault/ fatigue - cable overloading - malfunction of drawworks	Pipe assembly drops	-damage to rig floor	- R.I. & M.	2	4	1	7
				-damage to spider/ elevator units	-high Q.A. & control	2	4	0	6
				-damage to other equipment	-high factor of safety	2	3	1	6
				-loss of lowering pipe	-incl.overload safety alarm & tension limiter device.	2	3	2	7
				- injury/loss of life	-limit service life	2	1	2	5
G8-100	Drawworks does not operate	-control malfunction - equipment fault/ malfunction	Delay in operation	-	- R.I. & M. - high Q.A. and control	5	4	0	9
			: END :						

APPENDIX B

FAULT TREE ANALYSIS

EXTRACTS FROM CONTRIBUTION BY
SYSTEMS RELIABILITY SERVICE (SRS)
United Kingdom Atomic Authority
Culcheth, Warrington, U.K.
on

DRILLED EMPLACEMENT METHOD

ASSESSMENT OF RELEASE OF RADIOACTVITY TO THE ENVIRONMENT

CONTENTS

B 1 INTRODUCTION

The Systems Reliability Service of the United Kingdom Atomic Energy Authority were invited to assist in the preparation of this hazard assessment. They have examined fault trees provided by TWC, suggested alternative formats and have assessed the probable frequency of occurrence of the various failure modes which have been postulated. Their computer programme ALMONA has been used to analyse the overall fault tree and assess the sensitivity of the Top Event to variations in the failure rates of particular components in the logic.

B 2 ASSUMPTIONS MADE FOR FAULT AND DATA ANALYSIS

1. Equipment will be designed, manufactured, maintained and tested to the appropriate nuclear standards.

2. Recovery of sunken flasks or active materials will not be precluded on cost considerations.

3. The equipment will be fully tested under all likely operating conditions.

4. Hoists and winch will have upper limit and overload trips and will incorporate redundancy in braking systems; one brake will operate on the rope drum and be actuated by an overspeed trip. Loss of power will bring on the brakes.

5. Flasks will be designed to withstand the maximum thermal and dynamic conditions which can be envisaged.

6. Flask lifting features will be designed to obviate faulty attachment.

7. Decks on ship and platform will restrain flasks from falling overboard in tilt situations.

8. Canisters will be sufficiently robust to withstand falls and mishandling.

9. A fallen canister could be retrieved from any position in the wet cell, inspected and returned to the transit flask if damaged.

10. Stringer support pipes will be inspected after ech use. The effect of any damage caused by the jaws of the lowering jacks will be considered in the light of results from previous test programme.

11. Ample redundancy will be provided where the requirement is indicated by future, more detailed, reliability assessments.

12. Procedure safeguards/interlock arrangements will ensure that the probability of the Remote Disconnect Joint at the top of the Active Stringer being released prematurely, is very low.

13. Design features will reduce the possibility of transport flask falling free from the trolley if a fault situation arises whilst it is being winched up to the platform.

B 3 FAILURE LOGIC

The study and proposals from TWC are presented in broad outline with some detail where necessary for clarification. The failure logic has been developed to the same depth but has postulated some possible areas for component failure where this has been thought useful to the fuller understanding of the possibilities for sub-system failure: individual failure rates for such breakdown have not always been thought necessary.

Failure logic is presented in Figures B1 to 15.

B 4 FAILURE RATE PREDICTIONS

Some comments on the derivation of the predicted failure rates have been given in Appendix B1.

In general failure rates for listing equipment and for individual components have been derived from data contained in the SRS Data Bank and from previous SRS Reports on similar equipment, flask transporting ships and drilling platforms.

The North Sea Environmental Guide (Ref. B2) has provided details of casualties to platforms over an 18 year period. Ship casualty rates have been derived from a report compiled by Lloyds for the Ministry of Defence Salvage Dept. and British Nuclear Fuels and from a Hazard Report on Canvey Island Operations (Ref. B3).

Some subjective judgement has been used in the predictions for success in recovery of sunken flasks and active containers. Previous exercises to locate and recover equipment (including active materials) from the sea bed have given some confidence in this area.

The assessment has considered any loss of a loaded container as a release of radioactivity to the environment; this is on the assumption that corrosion will eventually allow a dissipation of the contents. It is probable that deterioration to such an extent would require exposure over many years, giving equivalent time scales for location and for development of recovery procedures. However, such long time scales have not been allowed in the predictions for the probability of successful recovery.

It has been assumed that the cost of a recovery operation would not be a prohibiting factor if there was a possibility of a hazardous release of radioactivity.

The general assumption has been that the disposal system would be fully tested with non-active loads and that the equipment is proved to be capable of withstanding the severe environmental conditions (given that the worst of these conditions will be avoided by operating restrictions and emergency measures).

B 5 COMPUTER ANALYSIS

The failure logic has been simulated by the ALMONA computer programme at the Systems Reliability Directorate. The effect of varying the failure predictions in some of the most sensitive areas has been given with the computer analysis.

B 6 COMMENTS

The study has indicated that the operations presenting the highest probability for the release of radioactivity are those involved with the lowering of the stringers of active canisters from the platform to the borehole.

The hazard here is that canisters will be released and that recovery will not be possible (see Figures B8 and B5, items 41 and 42).

The main basis for the prediction is that the water depth of 5500m necessitates approximately 400 length of pipe to support the stringers and an equal number of pipe joints and welds. These will be subjected to high and fluctuating, strains and stresses by the ocean currents and the vertical load. Increasing the design safety factor (e.g. by increasing the pipe wall thickness) may not be an effective solution as the vertical lod could then be unacceptable.

A further significant contribution to the predicted top event during the lowering operation comes from the possibility that the platform might drift from its position (by more than 240m), during the time that a canister string has entered the borehole but not to its fully lowered position. (Items 54 to 56). This could result in fracture or distortion by overstress to the lowering assembly. (This is somewhat pessimistic in its portrayal of the hazard situation as the active waste would be contained and at least partly buried).

Some thoughts have been given to the possibility of providing a redundant (secondary) support system for the stringer load, so that in the event of support pipe failure, the stringer could be hauled back to the platform. Such redundancy has not been included in the present exercise.

Ship sinkings are not seen to involve a high incidence of loss for transport flasks as the most probable areas for such events are in congested (shallow) waters where the probability of successful location and recovery is relatively high. (See Figure B2).

Transference of flasks from ship to platform (by means of winching a bogie up a connecting ramp) is not expected to be without incident but the probability of losing a flask overboard is predicted to be less than 10^{-4} per year. (Fig. B4 and Item 16). This prediction assumes that careful examination will be given to sensitive areas during detail design of the associated handling equipment.

Normal operating procedures should present a very low hazard to personnel involved. (Figure B10). Maintenance activities in the wet cell will require careful pre-planning and close management control.

B 7 CONCLUSIONS

The calculated probability for the release of radioactivity to the environment is 10^{-2} per year, (1 chance in 100 for each year of operation).

The major contribution to this rate comes from the possibility that some of the stringers containing radioactive waste will be lost during the operation of lowering them to the sea bed: it may be possible to devise a satisfactory secondary support system to reduce this possibility.

The assessment has been based on outline design proposal and deals with operations and recovery methods using a constantly expanding technology. Some degree of engineering judgement has been used in arriving at the predicted failure rates and undue reliance should therefore not be placed on individual figures.

B 8 REFERENCES

B.1 A Feasibility Study of the Disposal of Radioactive Waste in Deep Ocean Sediments. Report No. 014N/83/2539.

B.2 The North Sea Environmental Guide 91984). Oilfields Publications Ltd., Ledbury.

B.3 Canvey. An Investigation of Potential Hazards from Operations in the Canvey Island/Thurrock Area. Report Nos. IBN 011883200-K, IBN 011883459-2.

DATA ANALYSIS - EXPLANATORY NOTES

APPENDIX B1

LOGIC REFERENCE	DERIVATION OF FAILURE RATES OR PROBABILITY	ASSESSMENT (θ = f/year) (p = f probability)
1	From Figure B1 logic	$\theta = 1.02 \times 10^{-2}$
2	From Figure B2 logic	$\theta = 1.06 \times 10^{-3}$
3,4	The flask will be designed and tested to ensure that it will be able to withstand the maximum thermal and dynamic conditions to which it can possibly be subjected. Tests on similar transport flasks have indicated that no breach will occur under these conditions.	$\theta = 5 \times 10^{-6}$
5	From Figure B5 logic	$\theta = 8.4 \times 10^{-3}$
6	From Figure B9 logic	$\theta = 10^{-6}$
7	From Figure B10 logic	$\theta = 3.5 \times 10^{-6}$
8	From Figure B11 logic	$\theta = 4.5 \times 10^{-4}$
9	After the canister stringers have been loaded into the borehole, a liner must be removed from its upper end before the hole is backfilled. (This is necessary because its corrosion could allow a leak path). The liners will be 0.66 m. dia x 275 m. long tubes - probably aluminium. Methods for achieving this are available. As there will be plenty of time to complete this operation before a hazard arises, failure is seen as a remote possibility.	$\theta = 5 \times 10^{-6}$

LOGIC REFERENCE	DERIVATION OF FAILURE RATES OR PROBABILITY	ASSESSMENT (θ = f/year) (p = f probability)
10	The possibility that a flask could not be recovered from shallow water in the harbour is considered to be remote in view of successful recoveries of materials from much more difficult situations.	$\theta = 10^{-6}$
11	From Figure B3 logic	$\theta = 1.23 \times 10^{-3}$
12	It is anticipated that the flasks will incorporate a feature to assist location under water. It should therefore be possible to ascertain the position even if one has broken away from a sunken ship. The probability of failure to recover <u>all</u> flasks sunk with a transport ship is a function of failure to locate (feature failed or inefficient for various reasons) and failure to recover (impossible situation) before flasks were breached by corrosion. The water depth could be between 200 and 5500m. The possibility that the <u>cost</u> of a recovery could inhibit any action is not taken into account here. The probabilities are (somewhat subjectively) assessed as:- Failure to locate every flask 0.5 Failure to recover every flask after location 0.5 Failure to locate <u>OR</u> Recover = 1 - (0.5 x 0.5)	p = 0.75
13	Statistics provided by Lloyds Register indicate that 0.33% of the world's cargo fleets are lost annually as a result of being stranded or being involved in collisions, most of which would be in congested coastal waters. This figure covers a wide range of ships and navigating equipment.	$\theta = 10^{-3}$

LOGIC REFERENCE	DERIVATION OF FAILURE RATES OR PROBABILITY	ASSESSMENT (θ = f/year) (p = f probability)
14	Location/recovery would not be expected to be so difficult as in the deep water situation (ref. 11 above) but rocks and sea conditions could make access difficult. 100% success has been achieved in similar exercises.	$p = 2 \times 10^{-2}$
15	From Figure B4 logic	$\theta = 6.1 \times 10^{-4}$
16	This recovery differs from ref. 11 above in that only one flask is involved (and no ship). The approximate location would be known.	$p = 0.2$
17	The flasks will sit within a well deck or below deck. Collision with another ship or with other obstructions in coastal waters is not likely to cause a flask to fall overboard and the combined possibility is very low.	$\theta = 2 \times 10^{-4}$
18	The possibility of recovering a flask from relatively shallow coastal waters is seen as similar to ref. 14 above.	$p = 2 \times 10^{-2}$
19	Statistics provided by Lloyds Register indicate that 0.38% of the worlds cargo fleets are lost in the open sea each year. Most of these losses are due to foundering after hull failure, cargo shift or leakage through hatches or machinery pipelines. A large proportion of the craft are fishing vessels and small craft, a major contributory factor being neglectful maintenance. It is noted that very few vessels are actually sunk by fire damage. The vessel used for transporting flasks will be designed, built and maintained to high standards.	$\theta = 5 \times 10^{-4}$

LOGIC REFERENCE	DERIVATION OF FAILURE RATES OR PROBABILITY	ASSESSMENT (θ = f/year) (p = f probability)
20	From Figure B3 logic	$\theta = 7.3 \times 10^{-4}$
21	Statistics from Lloyds Register indicate an extremely low incidents of fire damage causing a ship to sink. The supply ship will not be carrying an inflammable cargo.	$\theta = 10^{-6}$
22	From Figure B3 logic	$\theta = 7.3 \times 10^{-4}$
23	The flasks (~150 each year) will be loaded on a trolley and hauled up a (100 m) bridge from ship to platform by a winch/cable system. A brake would operate from trolley to track in the event of a runaway situation. On the assumption that the winch will be designed to nuclear standards with an emergency brake on the rope drum and overspeed/overload trips the incidence of a runaway commencing is assessed as 2×10^{-3} per year. The trolley brake would reduce the possibility of a high speed impact on the ship to 4×10^{-4} per year. The probability that such an impact would cause sufficient damage to sink the ship is assessed as 0.3.	$\theta = 1.2 \times 10^{-4}$
24	An uncontrolled lowering of the transfer bridge caused by a failure on the pedestal crane could allow it to fall heavily on to the ship. Assuming design standards as for ref. 22, the incidence of crane failure or the operator allowing an uncontrolled fall is assessed as 10^{-2} per year. The probability that the bridge would strike the ship and cause sufficient damage to sink it is assessed as 0.01.	$\theta = 10^{-4}$

LOGIC REFERENCE	DERIVATION OF FAILURE RATES OR PROBABILITY	ASSESSMENT (θ = f/year) (p = f probability)
25	The pedestal crane will not normally extend over the supply ship. The combined incidence of a fault allowing it to fall at the same time as a positional drift of the ship is assessed as 10^{-5} per year. The probability that the impact would be sufficient to sink the ship is assessed as 0.05.	$\theta = < 10^{-6}$
26	Operations will not be conducted in winds above Beaufort 5 (wave heights over 1.5m). The ship is not required to approach closer to the platform than about 60m and will be fitted with station-keeping equipment. It will be in the vicinity of the platform for about 50 days per year. North Sea Platform Records 1966-1983 (say 2000 platform years) note 11 Collisions with ships; 7 of these were supply boats and 2 of these sank.	$\theta = 5 \times 10^{-4}$
27	In view of the precise manoeuvring required in order to engage the transport bridge with the ship, some collisions would be expected. However, the ship end of the bridge is free to move in 3 rotational and 1 linear direction and inertia forces only would be involved. The possibility that these could cause sufficient damage to sink the ship is judged to be remote.	$\theta = 10^{-5}$
28	The flasks are taken from their storage area to the transfer trolley by a portal crane. The incidence of an uncontrolled fall of the load either from a crane fault or by human error is assessed as 10^{-3} per year. However the flask need only be lifted to trolley height and it should be possible for the designers to ensure that such a fall has little likelihood of causing enough damage to sink the ship.	$\theta = 5 \times 10^{-6}$

LOGIC REFERENCE	DERIVATION OF FAILURE RATES OR PROBABILITY	ASSESSMENT (θ = f/year) (p = f probability)
29	The flasks will sit within a well deck or below deck; the possibility of a fall would then arise only from a collision or severe listing (leakage/ballasting). Ship sinkage and ship damage from collisions in coastal waters have been dealt with above; so that the assessment here is for incidents in the open sea which cause sufficient damage/listing for a flask to be cast overboard.No flask handling is required during transit.	$\theta = 10^{-6}$
30	The flasks will be handled by the portal crane within the confines of a well deck. The incidence of a fail completely free from the crane is assessed as 10^{-4} per year; the probability that it will then fall overboard is assessed as 0.01. (This assumes that the flask lifting features will be designed to obviate faulty attachment and that the flask will be of a shape which does not roll easily).	$\theta = 10^{-6}$
31	A runaway bogie completely free from the transfer winch implies a rope or attachment breakage followed by failure of the emergency brake on the bogie. The flask could also fall free from the bogie following impact (or rapid deceleration) after running away, (see ref. 23 above) and fall overboard.	$\theta = 5 \times 10^{-4}$
32	The transfer bridge will not be rigidly secured at the ship and will be free to pivot on the platform. The loading should therefore be determined with reasonable accuracy. High loading at the ship to bridge connection will cause the bridge to be lifted clear.	$\theta = 10^{-5}$

LOGIC REFERENCE	DERIVATION OF FAILURE RATES OR PROBABILITY	ASSESSMENT (θ = f/year) (p = f probability)
33	The pedestal crane will be used to transfer the flasks to a parking area on the platform and later through an opening from the deck into the wet cell. The incidence of a crane fault or human error which would allow a flask to fall free of any constraint is assessed as 10^{-2} per year. All movements will be within the deck of the platform. The probability that a flask would fall into the sea after breaking away is assessed as 10^{-2}.	$\theta = 10^{-4}$
34	The platform will be quite stable under normal conditions. There will be a barrier around the deck. A structural failure or fault in the ballasting system could not allow it to tilt sufficiently to allow a free standing flask to fall from the deck into the sea.	$\theta = 10^{-6}$
35	Two canister movements are required, from transport flask to storage rack and from there to a stringer pipe; in total about 5500 movements per annum. Operations will be carried out by a manipulator/grab carried on an overhead crane. The flask and canisters will be under water: positioning will be automatically programmed or manual with visual aid. Similar movements are carried out regularly in the nuclear industry and whilst the equipment is in regular need of maintenance both planned and breakdown, there is no evidence of failures which allow the load to fall. However, this may be because such incidents would be sensitive and information restricted unless a general hazard was involved.	

LOGIC REFERENCE	DERIVATION OF FAILURE RATES OR PROBABILITY	ASSESSMENT (θ = f/year) (p = f probability)
35 cont'd	Engagement/release of grab to canister is by downward pressure so that there is a minimal operator/controls function to give scope for errors which could allow the load to fall free. It is assumed that safety features will be provided to obviate any undue horizontal forces on the grab which might arise from operator or control system faults, and that there will be an overload trip on the hoist system. The assessment of a "drop" rate allows for some possible difficulties as the canisters are being loaded into the stringer (see Fig. B6)	$\theta = 0.411$
36	Operations do not require suspension of a canister directly above the moonpool but there is some possibility that a fallen canister could topple in that direction. The canisters are 0.43m dia x 1.3m long; the hole leading from the wet cell to sea through which the stringers pass is under 0.7m diameter; it is estimated that this hole will be closed for 95% of the canister movement time. A fabricated structure carrying the jacking (stringer lowering) mechanism will be permanently in position above the exit hole. The probability that dropped canister falling into the moonpool will then through a hole if open is assessed as 10^{-4}.	$p = 5 \times 10^{-6}$
37	An individual canister would be difficult to locate and recover, particularly if buried under mud on the sea bed.	$p = 0.9$

LOGIC REFERENCE	DERIVATION OF FAILURE RATES OR PROBABILITY	ASSESSMENT (θ = f/year) (p = f probability)
38	Stringer pipes are moved from their canister loading position to the grouting chambers and subsequently to the stringer assembly position by an overhead crane with a hook and lifting attachment. About 550 such moves will be required annually; however it is understood that during 95% of these moves the lm hole in the base of the moonpool will be filled and the jacking unit will always be positioned over the hole. The possibility of a pipe falling from the hoist and passing through the moonpool to sea is considered to be negligible. The stringer pipe assembly is moved downwards by a hydraulic jack system as the string is assembled. During this assembly time there are periods when the jack with its gripping spider is the only means of support for the stringer. However, there are two separate gripping mechanisms, one of which must be forced on to the pipe before the other can be released. The slips in the gripping jaws operate within a taper so that the weight of the pipe forces them into closer contact with its outside diameter. They can only be released by taking the weight of the pipe on the other set of gripping jaws. (See Fig.B7).	$\theta = 6 \times 10^{-5}$
39	In the event of the spider jaws failing open when the hoist is not attached, stringers must pass from the moonpool to sea.	p = 1.0

LOGIC REFERENCE	DERIVATION OF FAILURE RATES OR PROBABILITY	ASSESSMENT (θ = f/year) (p = f probability)
40	It is planned to fit sonar or other location devices to each stringer. No definite method of recovery has been established to date but the history of marine recovery operations gives grounds for optimism.	$p = 0.2$
41	From Figure B8 logic	$\theta = 0.042$
42	As for Ref. 40 - No particular difficulty is envisaged for removal of a damaged stringer from a borehole (refs. 55-57).	$p = 0.2$
43-46	See Reference 35 above	$\theta = 0.411$
47	It is estimated that a stringer being assembled will be attached to the hoist 70% of the time	$p = 0.3$
48-50	See Reference 38 above	$\theta = 2 \times 10^{-4}$
51	From Figure B8 logic	$\theta = 0.0405$
52	There will be roughly 400 pipe joints in the total length of stringer/support pipe. The design of joint has been well proven in the oil industry. The assessed failure rate assumes tht no active loads will be lowered until this joint has been fully tested: it includes failure of the pipe around the joint (including the weld).	

LOGIC REFERENCE	DERIVATION OF FAILURE RATES OR PROBABILITY	ASSESSMENT (θ = f/year) (p = f probability)
52 (cont)	The material used for the stringer support pipes has not been chosen; possibilities are steel, titanium and carbon fibre. Assuming testing of the design in working conditions and careful inspection during manufacture and after each operational use, a low possibility of failure is predicted.	$\theta = 0.03$
54-56	If the platform drifts significantly away from its station (~240m) whilst the stringer is being loaded into the borehole, high lateral loads could distort or shear the assembly. The incidence of such a platform malfunction during the time at risk (say 200 hours per year) is assessed as 0.05 per year. Emergency action would be taken to withdraw the stringer upon an indication of drift commencing. The possibility of such action being ineffective is assessed as 0.2.	$\theta = 0.01$
57	The remote disconnect joint is uncoupled to release the stringer from the support pipes once it has been lowered into position. Three separate actions are required, ensuring that the possibility of the joint being actuated prematurely by human error will be very low (assessed as 5×10^{-4}/y). The possibility of premature opening by electrical or mechanical faults during the time at risk (~ 100 hours per year) is assessed as 10^{-3}/year.	$\theta = 0.0015$

LOGIC REFERENCE	DERIVATION OF FAILURE RATES OR PROBABILITY	ASSESSMENT (θ = f/year) (p = f probability)
58	It is proposed that the pipe assembly will be lowered by means of a jacking system which clamps on to sections of the pipe as they are assembled and lowers by discrete motions of a hydraulic cylinder. At the point of entry into the borehole it might be necessary for the winch to take over in heavy swells, to compensate for the rise and fall of the platform. The possibility of a release of the load at any time is generally similar to that for the lowering system in the wet cell (Ref. 38 and Fig.B7).	$\theta = 2 \times 10^{-4}$
59-61	The atmosphere within a flask is sampled before the lid is removed. The flask would be returned without being opened if a breached canister was suspected. The canisters are quite robust (5mm wall thickness stainless steel). They are taken out of the flask and require two simple moves underwater within the shielded wet cell. Although there is a possibility of breaches caused by falls or mishandling, the possibility that a breach would be large enough to release any vitrified waste is judged to be remote. Minor activity released from any smaller breach would be insufficient to hazard the external environment. A fallen canister could be retrieved and would be returned to the transport flask if any damage was suspected	$\theta = 10^{-6}$

LOGIC REFERENCE	DERIVATION OF FAILURE RATES OR PROBABILITY	ASSESSMENT (θ = f/year) (p = f probability)
62-64	The transport ship and platform are dedicated to the handling of active waste containers. It is inconceivable that any personnel will be unaware of the dangerous contents of a flask. The tools and lifting equipment necessary for lid removal will only be provided within the wet cell.	$\theta = 10^{-6}$
65	The water in the wet cell will be interconnected with the sea and have the same surface level within 1.0m. A significant fall in cell water level could be caused if the water in two or more of the ballast containers was insufficient. The ballast tanks are compartmentalised; the possibility that a fault in level control system or pump operation would affect a large number of tanks to cause such an incident is remote. The predicted rate assumes that adequate redundancy will be provided as required by future reliability assessment.	$\theta = 2 \times 10^{-3}$
66	It is assumed that there will be normal and ultimate limit switches on the wet cell hoist. If the first limit is used as a stopping feature by the operator the over hoist frequency is assessed as 5×10^{-4}/y.	$\theta = 5 \times 10^{-4}$
67	It is assumed that the roof shield of the wet cell will normally be open only to receive or despatch flasks. A ballasting fault should be apparent long before that was a hazardous fall in the wet cell water level and ample warning could be given. It is assumed that an interlock will prevent wet cell operations proceeding until the roof shield is closed.	$\theta = 5 \times 10^{-3}$
68	See logic references 72-82	$\theta = 5 \times 10^{-4}$

LOGIC REFERENCE	DERIVATION OF FAILURE RATES OR PROBABILITY	ASSESSMENT (θ = f/year) (p = f probability)
69	A sunken platform would be reasonably easy to locate: access to flasks within the platform might be difficult, particularly if it rests at an awkward angle on the sea bed.	p = 0.9
70	From Figure B14 logic	$\theta = 3 \times 10^{-4}$
71	There will be some time between loss of station and an eventual possibility of stranding to make arrangements for the removal of radioactive material. It is assumed that plans and equipment will be available for all contingencies.	p = 0.01
72-82	Various possible causes have been postulated in Figs. 12 and 13 for sinkage of the platform; at this early stage in the design it would not be rewarding to examine these in any detail. In the records of casualties to North Sea Platforms during the years 1966 - 1983 (say 2000 platform years), 7 sinkings of platforms are noted - one whilst being towed; however it seems that most of these resulted from steel fractures whereas a concrete structure has been proposed for the disposal platform. It will be sited away from busy shipping lanes. Concrete structures have been performing satisfactorily; a particular hazard for the proposed platform is that a flask might fall overboard on to a buoyancy chamber. However, these will be compartmented and the predicted drop rate is not significant (ref. 33). Given that detailed reliability studies will be carried out on sensitive	

LOGIC REFERENCE	DERIVATION OF FAILURE RATES OR PROBABILITY	ASSESSMENT (θ = f/year) (p = f probability)
72-82 cont'd	areas such as, ballast and pumping systems the probability that it will be sunk whilst carrying radioactive materials is assessed as 5×10^{-4} per year.	$\theta = 5 \times 10^{-4}$
83	From Figure B14 logic	$\theta = 0.6$
84	The platform will be located about 1000Km from the nearest shoreline. In the event of a loss of position as a result of faults listed in Figures B14 and B15, there would be some time available before the platform was stranded either to correct the fault or to summon towing assistance: this time would obviously depend upon the direction and speed of drift, but must be a minimum of some days, giving time to correct most of the faults which might arise. A tow could be requested as a last resource.	$p = 5 \times 10^{-4}$
85	Possibilities for failure of the DP system are given in Figure B15 logic. At this early stage in the design it would not be rewarding to examine these in any detail. The overall failure rate of the system which is predicted, assumes that some redundancy will be provided in critical areas.	$\theta = 0.3$
86	The design/type of positioning power unit has not been developed in detail but some redundancy will be provided - the power has of course to be available at all times but full capacity is only needed in storm conditions. The failure rate predicted is indicated by an assessment of a diesel propulsion unit which used data from various sources including Lloyds Register.	$\theta = 0.2$

LOGIC REFERENCE	DERIVATION OF FAILURE RATES OR PROBABILITY	ASSESSMENT (θ = f/year) (p = f probability)
87	The predicted failure rates leading to loss of power to the DP System (refs. 88-90) are indicated by the previous work referred to for Ref. 86 above. The assessment given here could probably be much reduced by the provision of ample redundancy	$\theta = 0.1$
91	A fire or explosion could result in a prolonged loss of the positioning system, but no particular hazards can be seen.	$\theta = 10^{-3}$
92-94	See Ref. 85 above.	
95-101	See Ref. 85 above	

Fault Tree — Figure B1

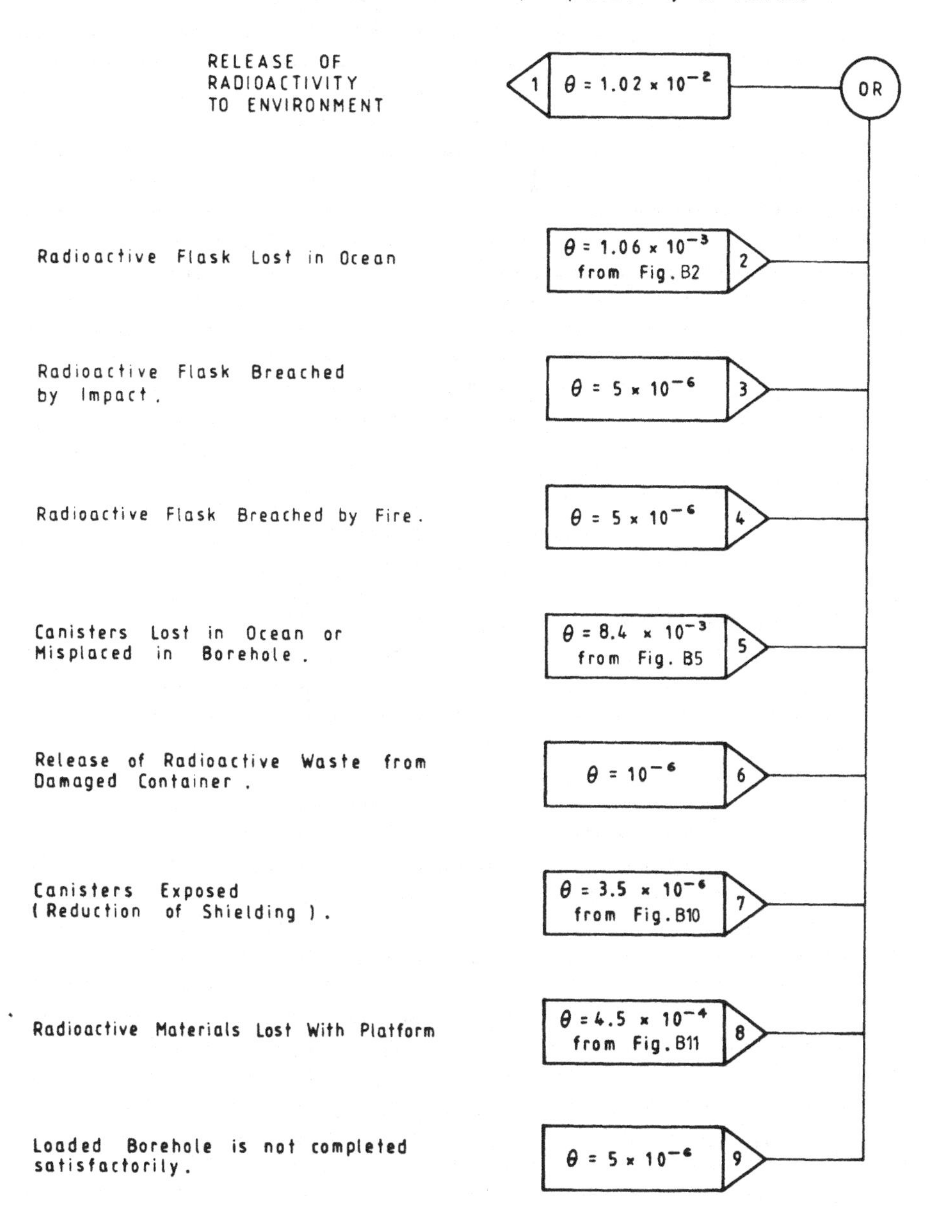

Fault Tree — Figure B2

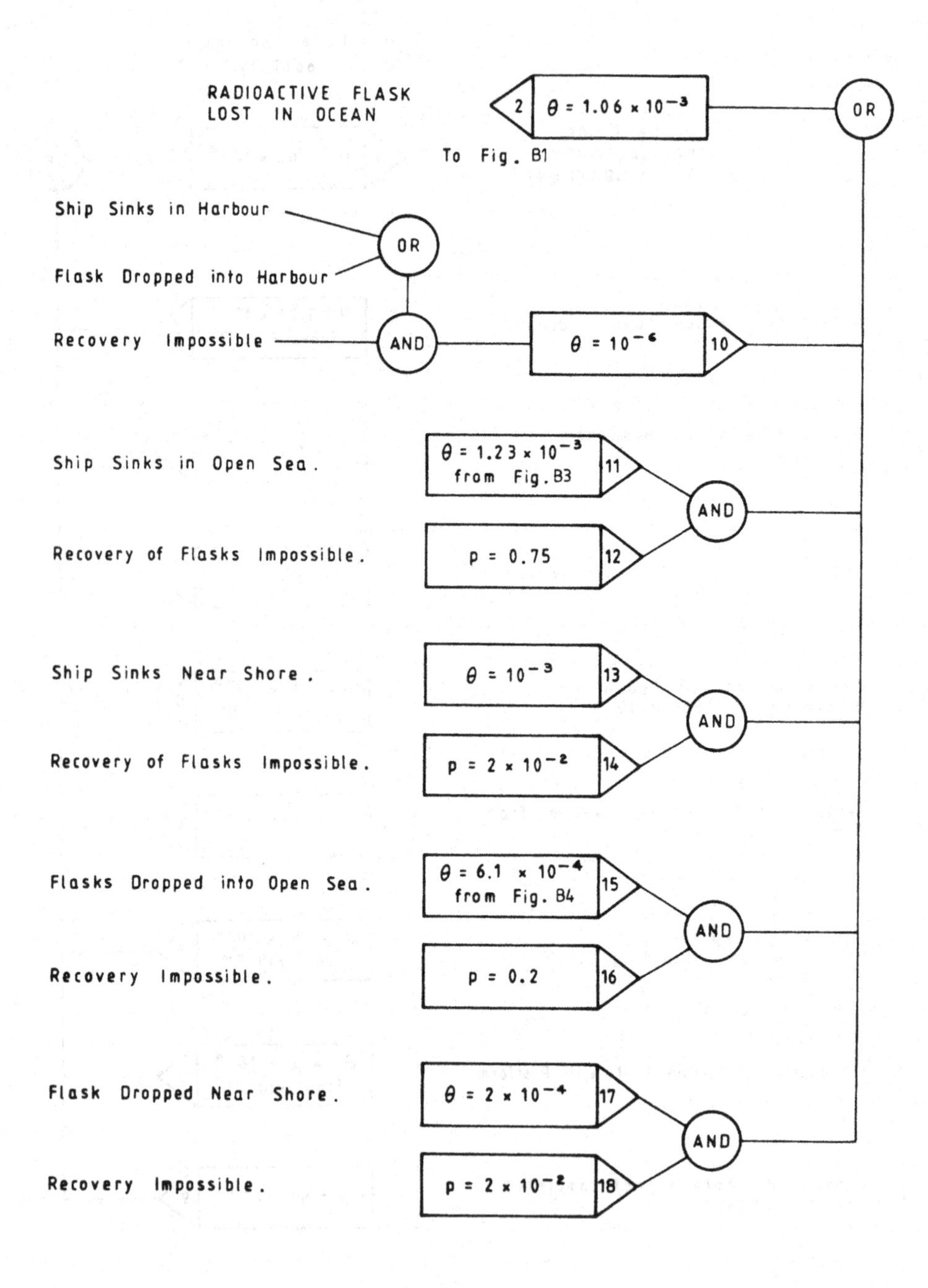

Fault Tree — Figure B3

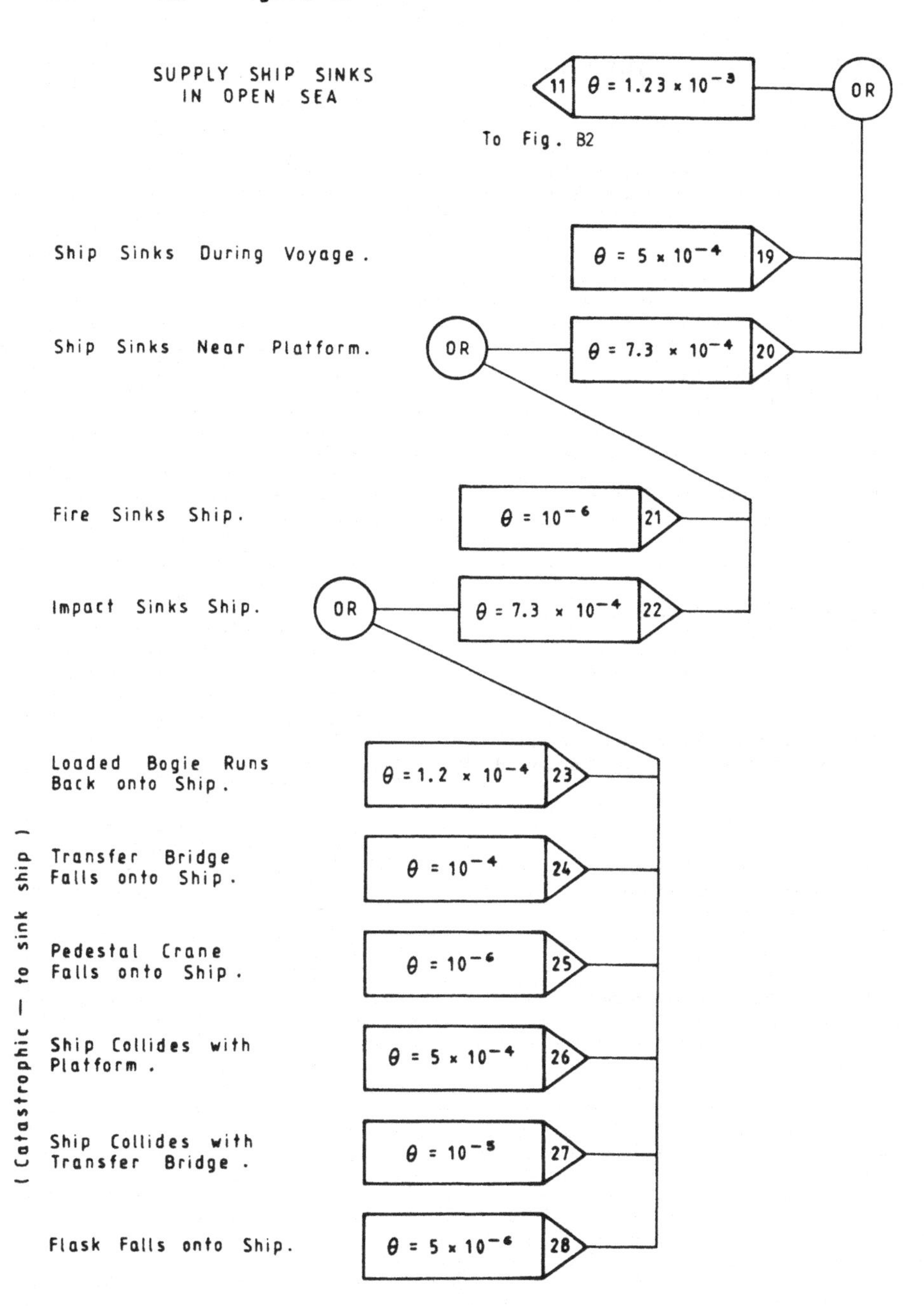

Fault Tree - Figure B4

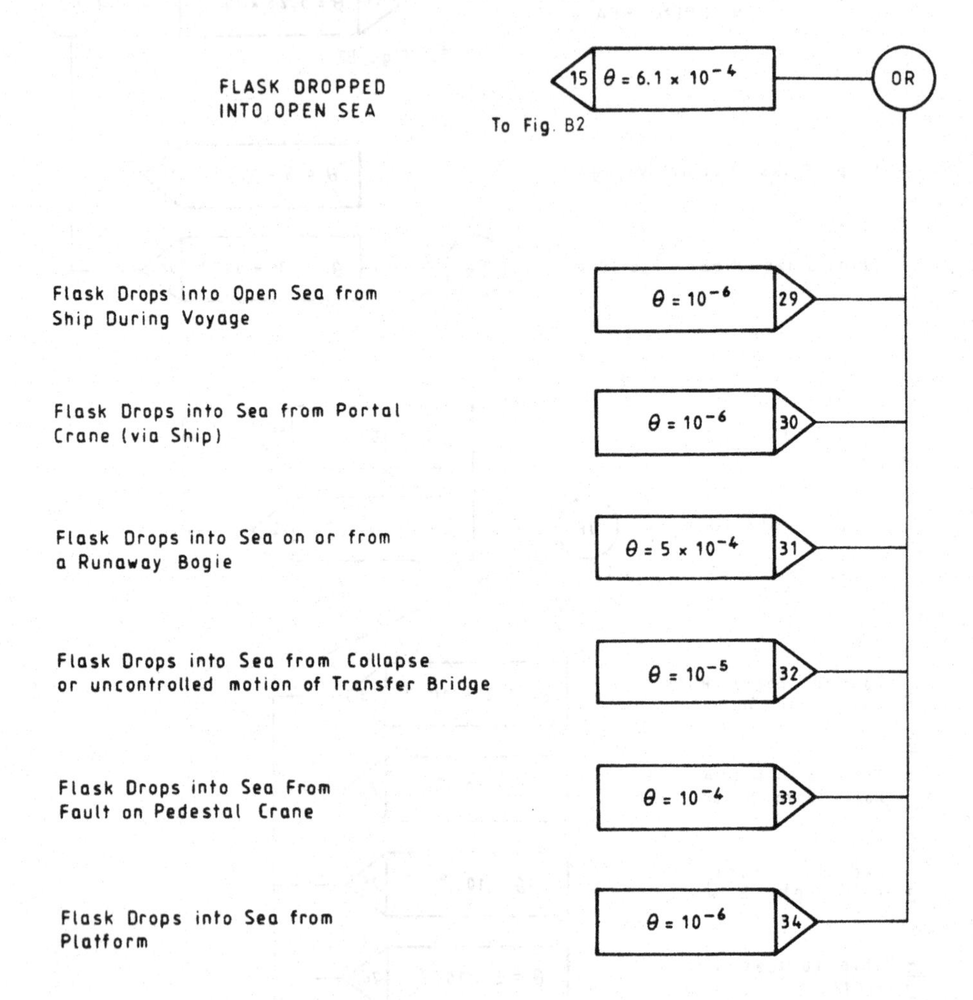

Fault Tree - Figure B5

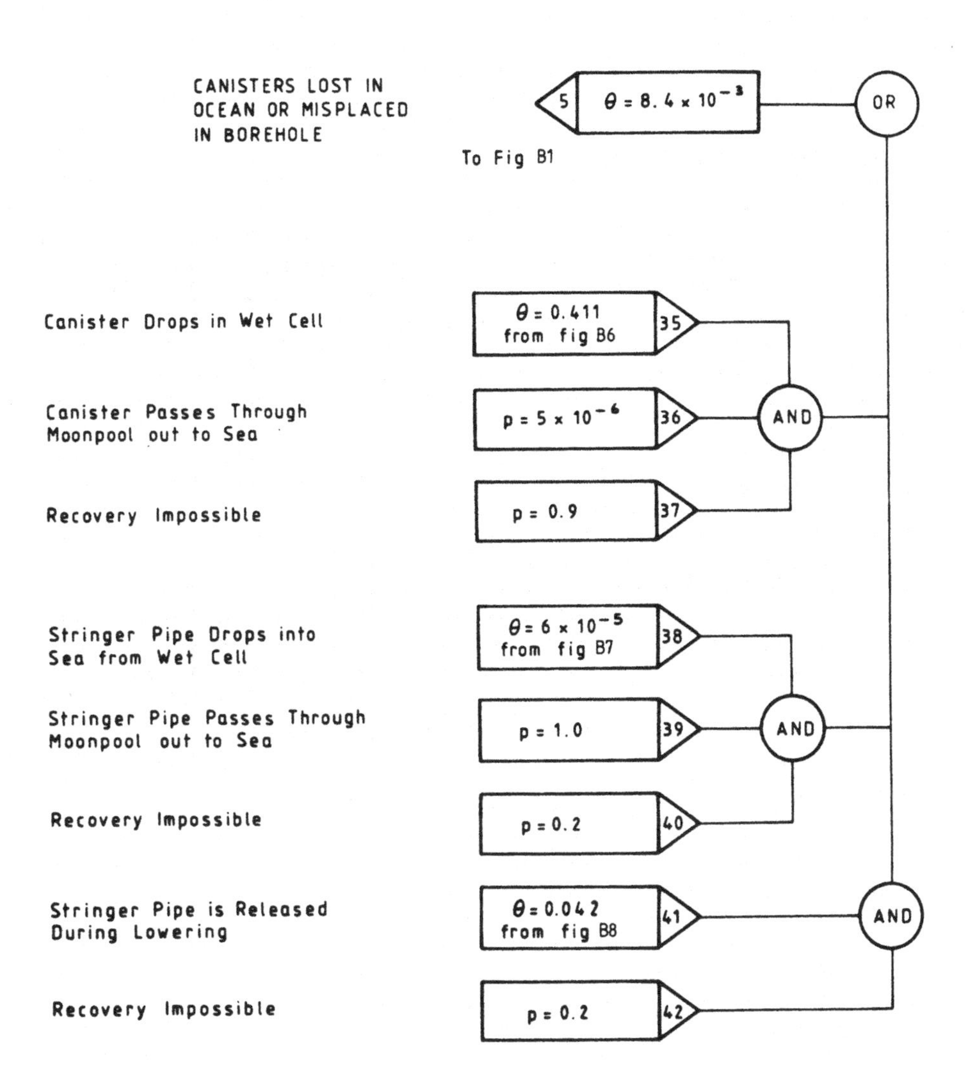

Fault Tree - Figure B6

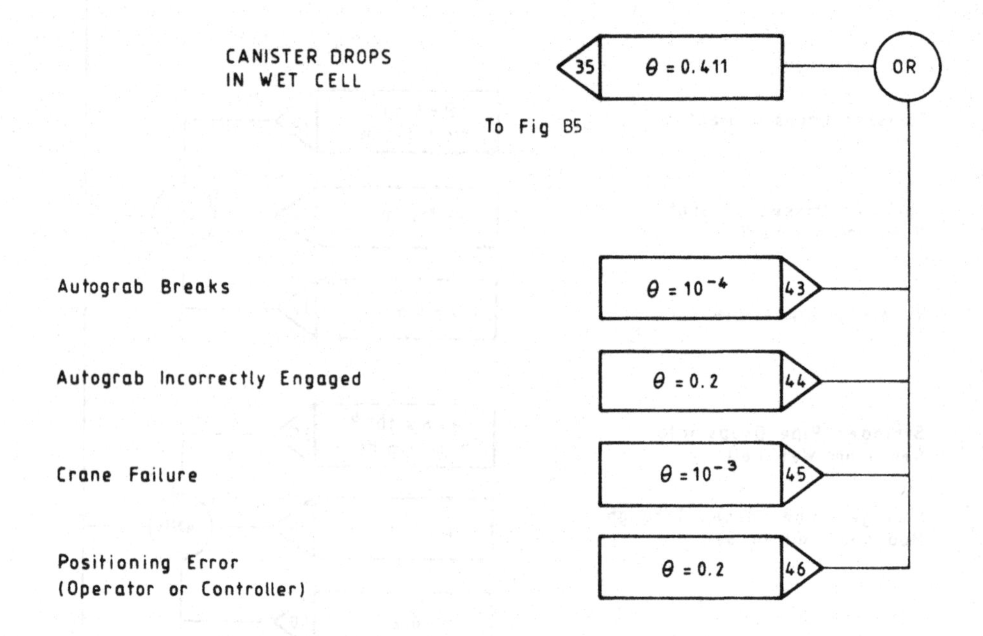

Fault Tree - Figure B7

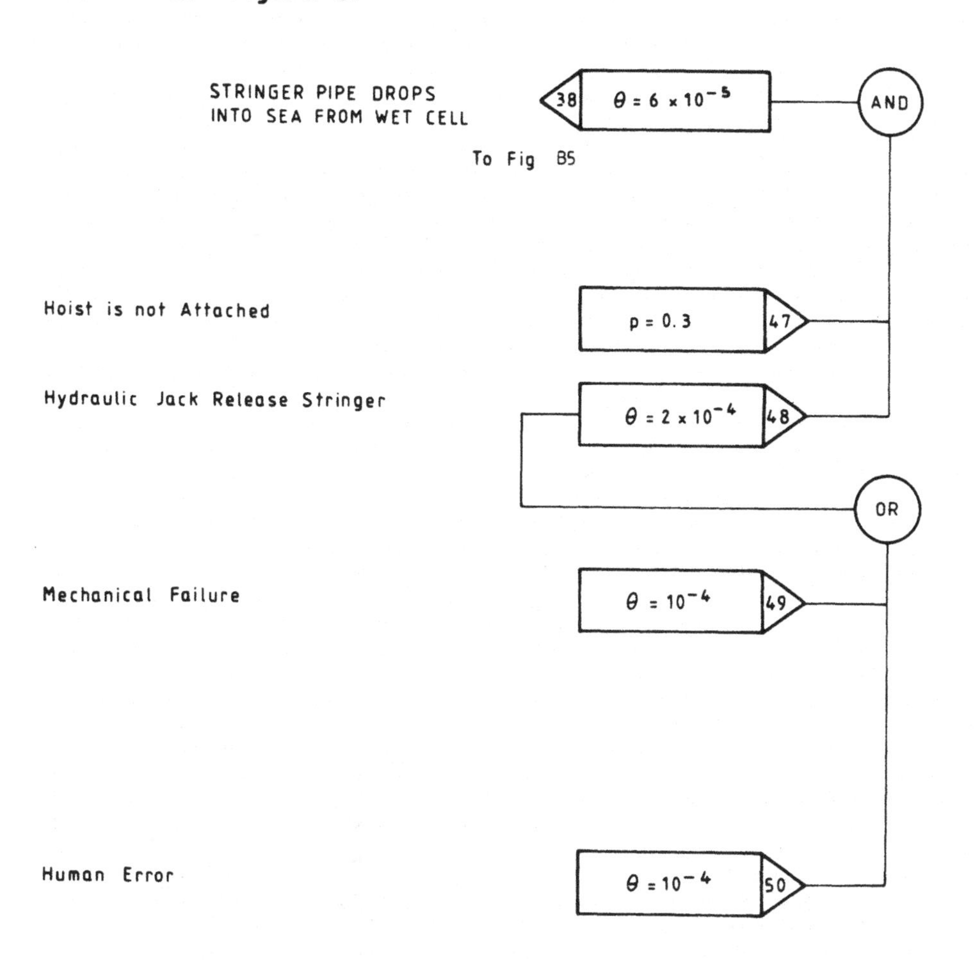

Fault Tree — Figure B8

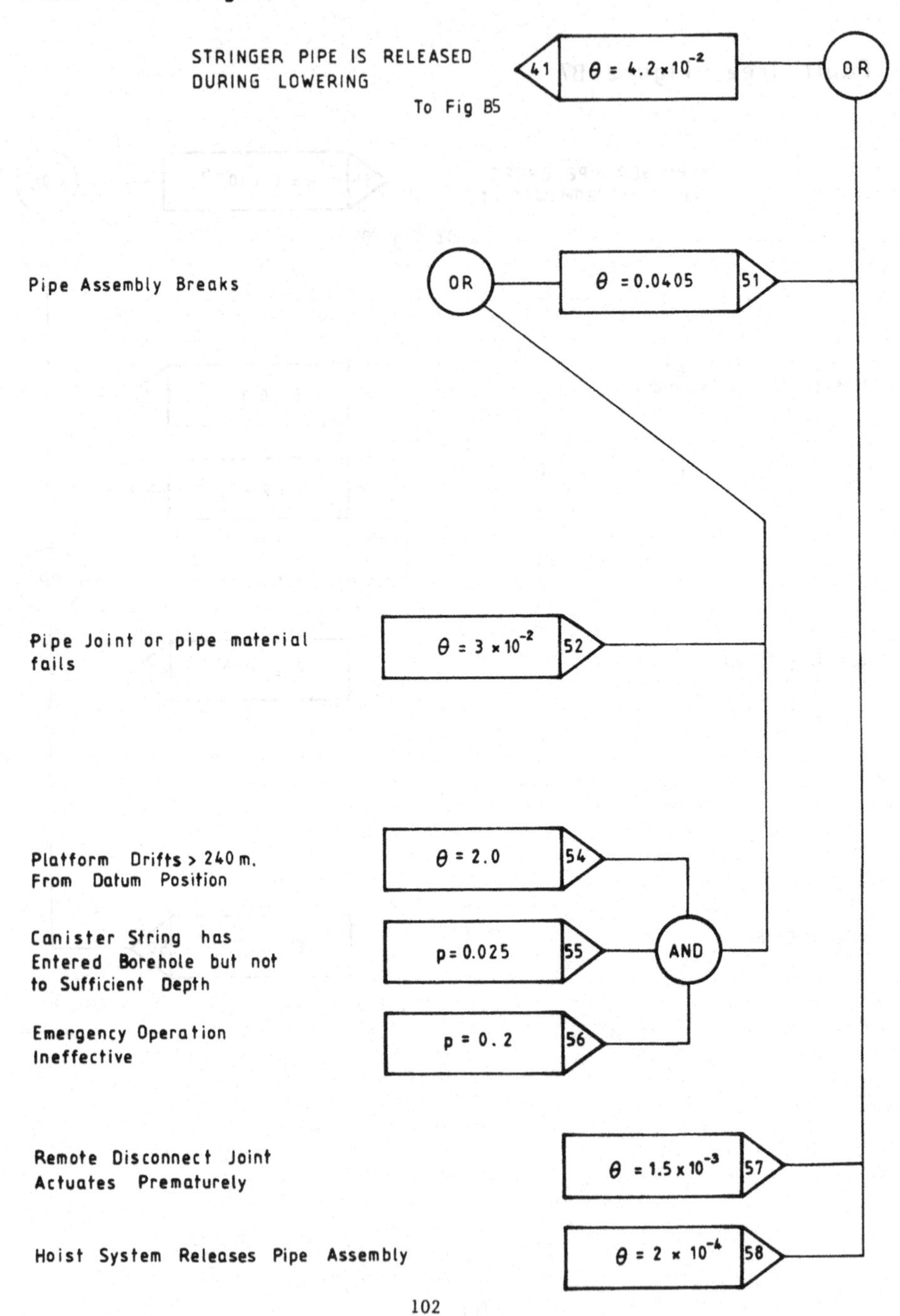

Fault Tree — Figure B9

RELEASE OF RADIOACTIVE WASTE FROM DAMAGED CONTAINER

Radioactive Waste Released to Environment

Canisters Breached Whilst Within Flask

Canisters Breached During handling

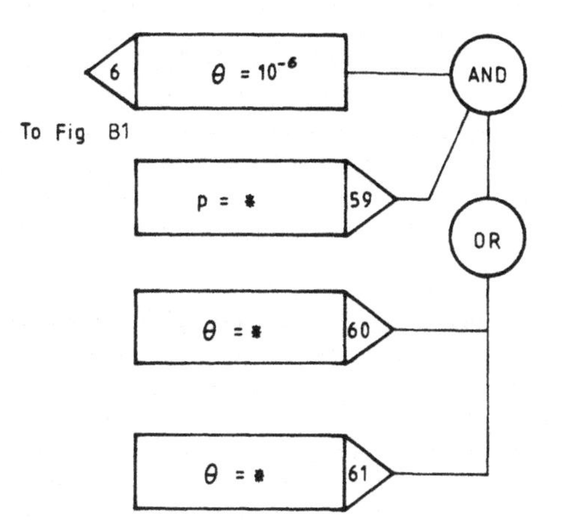

* See Derivation Notes Items 59 - 61

Fault Tree — Figure B10

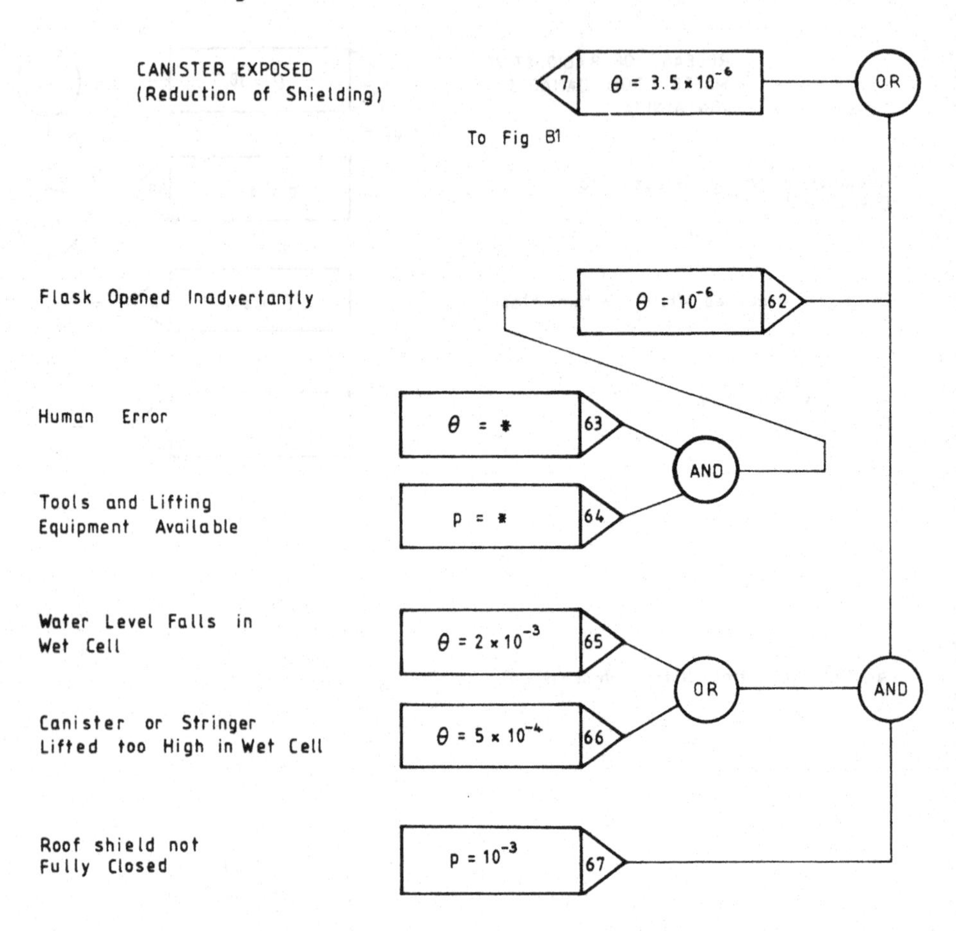

* See Derivation Note

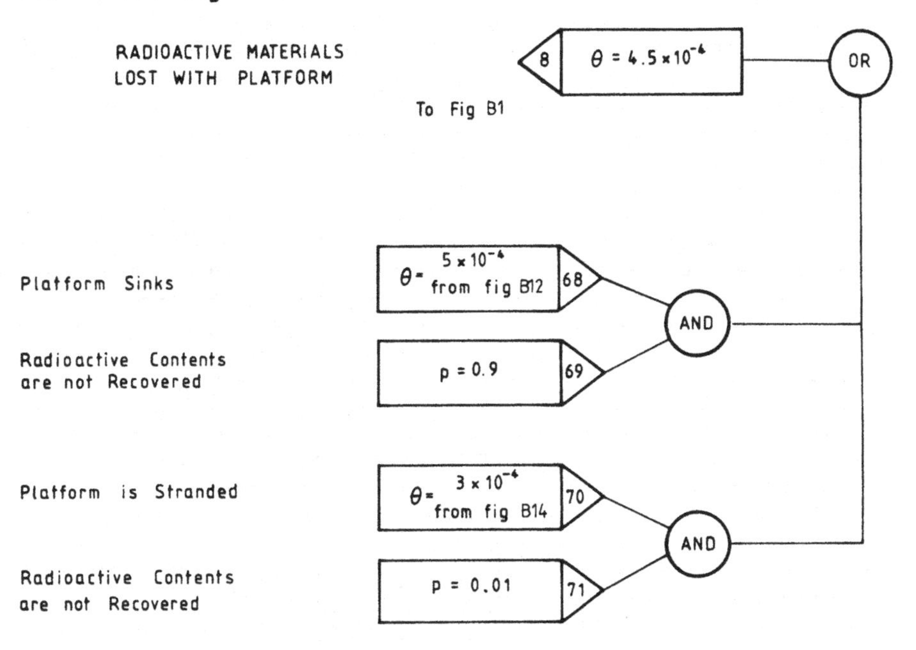
Fault Tree - Figure B11
RADIOACTIVE MATERIALS
LOST WITH PLATFORM
8
θ = 4.5 × 10⁻⁴
OR
To Fig B1
Platform Sinks
θ = 5 × 10⁻⁴ from fig B12
68
AND
Radioactive Contents are not Recovered
p = 0.9
69
Platform is Stranded
θ = 3 × 10⁻⁴ from fig B14
70
AND
Radioactive Contents are not Recovered
p = 0.01
71

Fault Tree — Figure B12

Event	Box	No.
PLATFORM SINKS	$\theta = 5 \times 10^{-4}$	68 (To Fig B11)
Platform Becomes Unstable	θ = *	72
Structure Fails under High Stress	θ = *	73
Ballast System becomes Faulty	θ = *	74
Fire Breaks Out	θ = *	75
Platform is Flooded	θ = *	76
Leak Occurs	θ = * from fig B13	77
Pumps Fail to Start	p = *	78
Pumps are unable to cope	p = *	79

Gates: 68 = OR (72, 76); 72 = OR (73, 74, 75); 76 = AND (77, OR (78, 79))

* See Derivation Notes

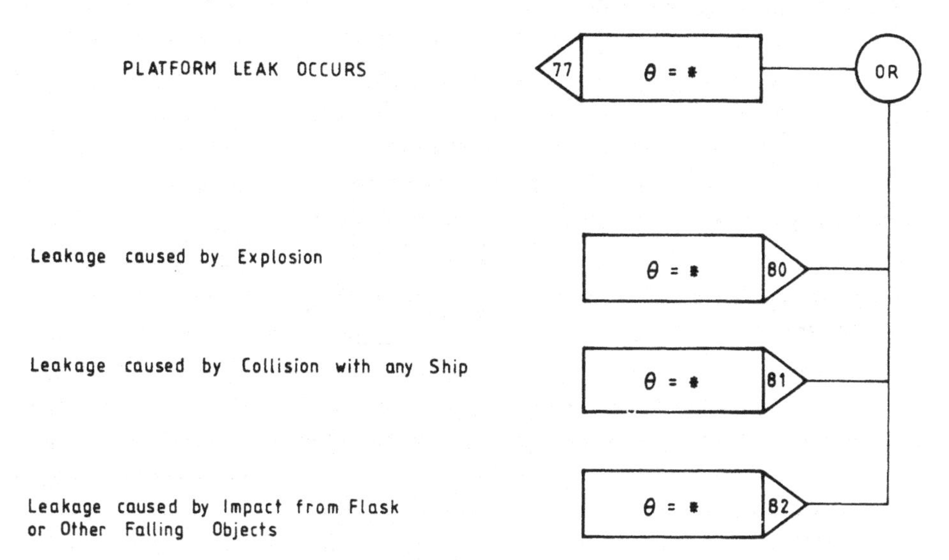
Fault Tree – Figure B13
PLATFORM LEAK OCCURS
77
θ = *
OR
Leakage caused by Explosion
θ = *
80
Leakage caused by Collision with any Ship
θ = *
81
Leakage caused by Impact from Flask or Other Falling Objects
θ = *
82
* See Derivation Notes

Fault Tree – Figure B14

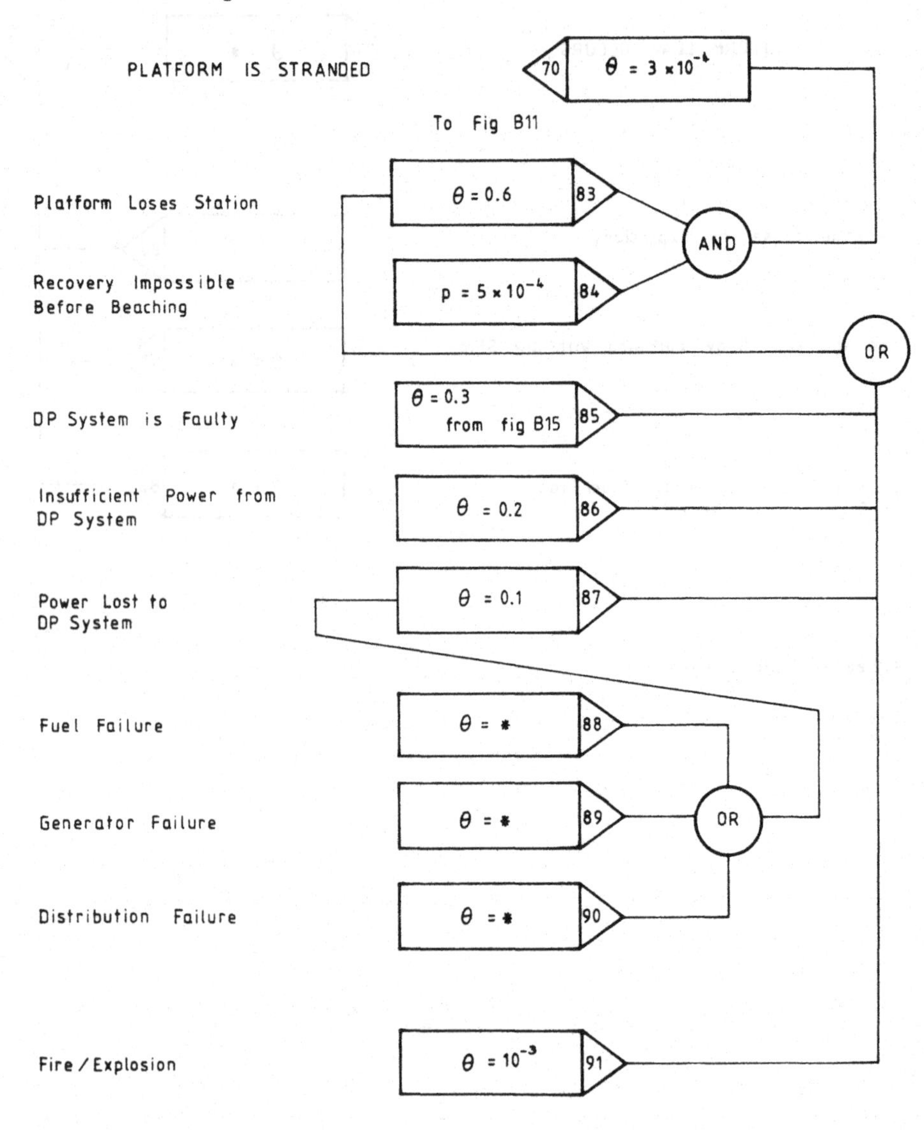

* See Derivation Notes

Fault Tree — Figure B15

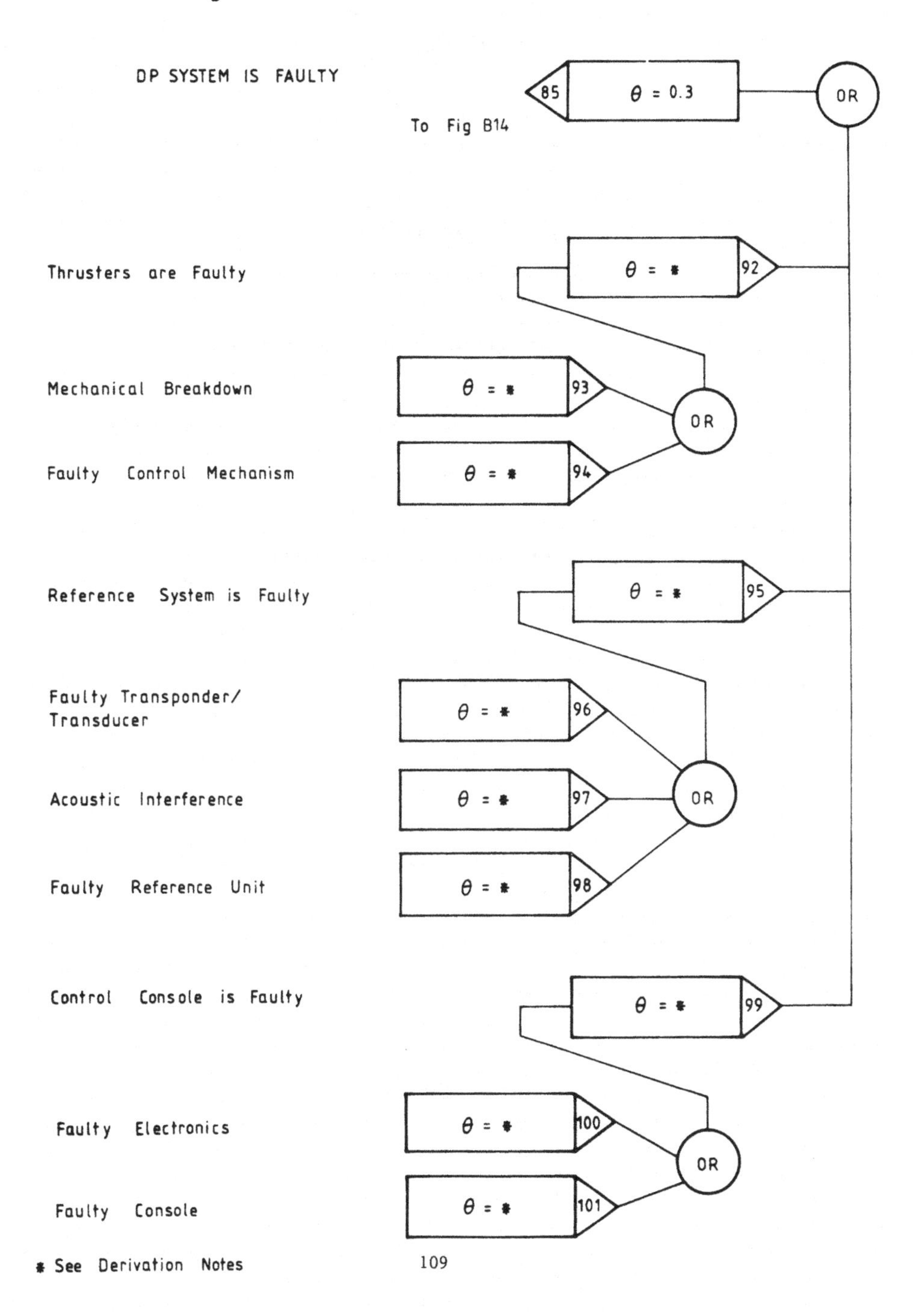

* See Derivation Notes

APPENDIX C

FAULT TREE ANALYSIS

SUMMARY OF COMPUTER OUTPUT
(T.W PROGRAM "CUTS")

DRILLED EMPLACEMENT METHOD
F.T.A. TOP EVENT:

Release of Radioactivity to the Environment

CONTENTS

APPENDIX C EXPLANATORY NOTES

DERIVATION OF IMPORTANCE VALUE

The fault tree analysis (F.T.A.) using Boolian algebra, results in t enumber of "minimum cut sets" each of which consists of the product of a number of basic events. The overall probability value is computed by adding the probability values obtained for each "minimum cut set". This method is approximate and is based on the approximation.

$$P\ (A + B) = P\ (A) + P(B)$$

when (A) and (B) are mutually exclusive events or the probability values are low enough to justify this approximation.

The importance value "I", of a basic event is the sum of the probability values of all minimum cut sets which contain the basic event, divided by the overall probability (approximately, the sum of the probability values for all minimum cut sets).

i.e. $I_A = P_A / Po$

where I_A = importance value for event "A"
P_A = the sum of the probability values for all minimum cut sets with event "A" included.
P_o = sum of the probability value for all minimum cut sets

$$P_o = P_A + P_B + P_C + \ldots$$

where P_A, P_B, P_C etc. are the probability values for each minimum cut set.

If the probability value for an event "A" is changed to a new value by a factor n,

$$P_1 = nP_A + P_B + P_C$$

The original probability value was:

$$Po = P_A + P_B + P_C$$

The change in the overall probability is therefore:

$$P_1 - Po = (n - 1)\ P_A$$

or

$$P_1 - Po \left[1 + \ 1 + (n - 1)\ PA / Po \right]$$

But $\frac{PA}{Po} = I_A$

$$P_1 = Po \left[1 + I_A\ (n - 1) \right]$$

P_1 being the new overall probability value
and P_o being the original probability value.

As an example, if the probability value for an event "A" is doubled (i.e. n = 2,

$$P_1 = Po\ (1 + I_A)$$

i.e. the overall probability value is increased by (I_A x Po).

CASE 3

DEEP DRILLED EMPLACEMENT - NO RECOVERY FROM DEEP WATER

FAULTS TREE WITH MAGNITUDE INCLUDED
Computer Program "CUTS"

Introductory Notes

The magnitude is included in the fault tree analysis by multiplying the probability values for events which lead to loss of canisters by a factor f, equal to:

f = canisters lost per event/total annual number of canisters.

The f values are as follows:-

- Cases related to loss of flasks, i.e. events A10, A15, A17, A3 : f = 0.01
- Cases related to ship sinking, i.e A13, A19, A21, A22, A85, A86, A87, A91, A9 and A4 (fire, breaching of flasks) and A68 (platform sinking) : f = 0.1
- Events A6, A35, A62 : f = 4 E-4
- Events A65, A66 : f = 4 E-3
- Events related to loss of a string of canisters, i.e. A49, A58, A57, A52, A54 : f = 0.02

DEEP DRILLED EMPLACEMENT
CASE 1 - SOME RECOVERY FROM DEEP WATER

FAULT TREE

25 gates, 40 basic events.

```
A1                          RELEASE OF RADIOACTIVITY
       due to: A2               RADIOACTIVE FLASK LOST IN OCEAN
           OR  A3               FLASK BREACHED BY IMPACT
           OR  A4               FLASK BREACHED BY FIRE
           OR  A5               CANISTERS LOST
           OR  A6               WASTE RELEASED FROM DAMAGED CONTAINER
           OR  A7               CANISTER EXPOSED
           OR  A8               WASTE LOST WITH PLATFORM
           OR  A9               BOREHOLE NOT COMPLETED SATISFACTORILY

A2                          RADIOACTIVE FLASK LOST IN OCEAN
       due to: A10              FLASK IN HARBOUR NOT RECOVERED
           OR  A111             DUMMY
           OR  A112             DUMMY
           OR  A113             DUMMY
           OR  A114             DUMMY

A111                        DUMMY
       due to: A11              SUPPLY SHIP SINKS IN OPEN SEA
          AND  A12              RECOVERY OF FLASKS IMPOSSIBLE

A11                         SUPPLY SHIP SINKS IN OPEN SEA
       due to: A19              SHIP SINKS DURING VOYAGE
           OR  A20              SHIP SINKS NEAR PLATFORM

A112                        DUMMY
       due to: A13              SHIP SINKS NEAR SHORE
          AND  A14              RECOVERY OF FLASKS IMPOSSIBLE

A113                        DUMMY
       due to: A15              FLASK DROPPED IN OPEN SEA
          AND  A16              RECOVERY OF FLASKS IMPOSSIBLE

A114                        DUMMY
       due to: A17              FLASK DROPPED NEAR SHORE
          AND  A18              RECOVERY OF FLASK IMPOSSIBLE

A20                         SHIP SINKS NEAR PLATFORM
       due to: A21              FIRE SINKS SHIP
           OR  A22              IMPACT SINKS SHIP

A5                          CANISTERS LOST
       due to: A115             DUMMY
           OR  A116             DUMMY
           OR  A117             DUMMY

A115                        DUMMY
       due to: A35              CANISTER DROPS IN WET CELL
          AND  A36              CANISTER SLIPS THRO' MOONPOOL
          AND  A37              CANISTER RECOVERY IMPOSSIBLE
```

DEEP DRILLED EMPLACEMENT
CASE 1 - SOME RECOVERY FROM DEEP WATER

A116 DUMMY
due to: A38 STRINGER PIPE DROPS THRO' MOONPOOL
AND A4Ø STRINGER RECOVERY IMPOSSIBLE

A38 STRINGER PIPE DROPS THRO' MOONPOOL
due to: A47 HOIST IS NOT ATTACHED
AND A48 HYDRAULIC JACK RELEASES STRINGER

A48 HYDRAULIC JACK RELEASES STRINGER
due to: A49 MECHANICAL FAILURE
OR A5Ø HUMAN ERROR

A117 DUMMY
due to: A41 STRINGER PIPE RELEASED DURING LOWERING
AND A42 STRINGER RECOVERY IMPOSSIBLE

A41 STRINGER PIPE RELEASED DURING LOWERING
due to: A58 HOIST SYSTEM RELEASES PIPE ASSEMBLY
OR A57 DISCONNECTS PREMATURELY
OR A51 PIPE ASSEMBLY BREAKS

A51 PIPE ASSEMBLY BREAKS
due to: A52 PIPE FAILS
OR A118 DUMMY

A118 DUMMY
due to: A54 PLATFORM DRIFTS WITH STRING IN HOLE
AND A56 EMERGENCY OPERATION INEFFECTIVE

A7 CANISTER EXPOSED
due to: A62 FLASK OPENED INADVERTENTLY
OR A12Ø DUMMY

A12Ø DUMMY
due to: A67 ROOF SHIELD NOT FULLY CLOSED
AND A121 DUMMY

A121 DUMMY
due to: A65 WATER LEVEL FALLS IN WET CELL
OR A66 CANISTER OR PIPE LIFTED TOO HIGH

A8 WASTE LOST WITH PLATFORM
due to: A122 DUMMY
OR A123 DUMMY

A122 DUMMY
due to: A68 PLATFORM SINKS
AND A69 WASTE NOT RECOVERED

A123 DUMMY
due to: A7Ø PLATFORM IS STRANDED
AND A71 WASTE NOT RECOVERED

DEEP DRILLED EMPLACEMENT
CASE 1 - SOME RECOVERY FROM DEEP WATER

A7Ø PLATFORM IS STRANDED
due to: A83 PLATFORM LOSES STATION
AND A84 RECOVERY IMPOSSIBLE BEFORE BEACHING

A83 PLATFORM LOSES STATION
due to: A85 DP SYSTEM FAULTY
OR A86 INSUFFICIENT POWER FROM THRUSTERS
OR A87 POWER LOST TO THRUSTERS
OR A91 FIRE/EXPLOSION

DEEP DRILLED EMPLACEMENT

CASE 1 - SOME RECOVERY FROM DEEP WATER

	BASIC EVENTS	FAILURE RATE
A10	FLASK IN HARBOUR NOT RECOVERED	0.100E-05
A12	RECOVERY OF FLASKS IMPOSSIBLE	0.750E 00
A13	SHIP SINKS NEAR SHORE	0.100E-02
A14	RECOVERY OF FLASKS IMPOSSIBLE	0.200E-01
A15	FLASK DROPPED IN OPEN SEA	0.610E-03
A16	RECOVERY OF FLASKS IMPOSSIBLE	0.200E 00
A17	FLASK DROPPED NEAR SHORE	0.200E-03
A18	RECOVERY OF FLASK IMPOSSIBLE	0.200E-01
A19	SHIP SINKS DURING VOYAGE	0.500E-03
A21	FIRE SINKS SHIP	0.100E-05
A22	IMPACT SINKS SHIP	0.730E-03
A3	FLASK BREACHED BY IMPACT	0.500E-05
A4	FLASK BREACHED BY FIRE	0.500E-05
A35	CANISTER DROPS IN WET CELL	0.410E 00
A36	CANISTER SLIPS THRO' MOONPOOL	0.500E-05
A37	CANISTER RECOVERY IMPOSSIBLE	0.900E 00
A40	STRINGER RECOVERY IMPOSSIBLE	0.200E 00
A47	HOIST IS NOT ATTACHED	0.300E 00
A49	MECHANICAL FAILURE	0.100E-03
A50	HUMAN ERROR	0.100E-03
A42	STRINGER RECOVERY IMPOSSIBLE	0.200E 00
A58	HOIST SYSTEM RELEASES PIPE ASSEMBLY	0.200E-03
A57	DISCONNECTS PREMATURELY	0.150E-02
A52	PIPE FAILS	0.300E-01
A54	PLATFORM DRIFTS WITH STRING IN HOLE	0.500E-01
A56	EMERGENCY OPERATION INEFFECTIVE	0.200E 00
A6	WASTE RELEASED FROM DAMAGED CONTAINER	0.100E-05
A62	FLASK OPENED INADVERTENTLY	0.100E-05
A67	ROOF SHIELD NOT FULLY CLOSED	0.100E-02
A65	WATER LEVEL FALLS IN WET CELL	0.200E-02
A66	CANISTER OR PIPE LIFTED TOO HIGH	0.500E-03
A68	PLATFORM SINKS	0.500E-03
A69	WASTE NOT RECOVERED	0.900E 00
A71	WASTE NOT RECOVERED	0.100E-01
A84	RECOVERY IMPOSSIBLE BEFORE BEACHING	0.500E-03
A85	DP SYSTEM FAULTY	0.300E 00
A86	INSUFFICIENT POWER FROM THRUSTERS	0.200E 00
A87	POWER LOST TO THRUSTERS	0.100E 00
A91	FIRE/EXPLOSION	0.100E-02
A9	BOREHOLE NOT COMPLETED SATISFACTORILY	0.500E-05

Probability limit = 0.000E 00 Maximum length of cut set = 10

DEEP DRILLED EMPLACEMENT

CASE 1 - SOME RECOVERY FROM DEEP WATER

Cut set listing for A1 : RELEASE OF RADIOACTIVITY
26 minimal cut sets

Probability	Cut set list		
Ø.6ØØE-Ø2	A42	A52	
Ø.2ØØE-Ø2	A42	A54	A56
Ø.547E-Ø3	A12	A22	
Ø.45ØE-Ø3	A68	A69	
Ø.375E-Ø3	A12	A19	
Ø.3ØØE-Ø3	A42	A57	
Ø.122E-Ø3	A15	A16	
Ø.4ØØE-Ø4	A42	A58	
Ø.2ØØE-Ø4	A13	A14	
Ø.6ØØE-Ø5	A4Ø	A47	A5Ø
Ø.6ØØE-Ø5	A4Ø	A47	A49
Ø.5ØØE-Ø5	A4		
Ø.5ØØE-Ø5	A3		
Ø.5ØØE-Ø5	A9		
Ø.4ØØE-Ø5	A17	A18	
Ø.2ØØE-Ø5	A67	A65	
Ø.184E-Ø5	A35	A36	A37
Ø.15ØE-Ø5	A71	A84	A85
Ø.1ØØE-Ø5	A6		
Ø.1ØØE-Ø5	A62		
Ø.1ØØE-Ø5	A1Ø		
Ø.1ØØE-Ø5	A71	A84	A86
Ø.75ØE-Ø6	A12	A21	
Ø.5ØØE-Ø6	A71	A84	A87
Ø.5ØØE-Ø6	A67	A66	
Ø.5ØØE-Ø8	A71	A84	A91

OVERALL PROBABILITY OF FAILURE = Ø.987E-Ø2

DEEP DRILLED EMPLACEMENT

CASE 1 - SOME RECOVERY FROM DEEP WATER

ANALYSIS OF IMPORTANCE

	Basic Event	Failure Rate	Importance
A42	STRINGER RECOVERY IMPOSSIBLE	Ø.2ØØE ØØ	Ø.844E ØØ
A52	PIPE FAILS	Ø.3ØØE-Ø1	Ø.6Ø8E ØØ
A54	PLATFORM DRIFTS WITH STRING IN HOLE	Ø.5ØØE-Ø1	Ø.2Ø3E ØØ
A56	EMERGENCY OPERATION INEFFECTIVE	Ø.2ØØE ØØ	Ø.2Ø3E ØØ
A12	RECOVERY OF FLASKS IMPOSSIBLE	Ø.75ØE ØØ	Ø.935E-Ø1
A22	IMPACT SINKS SHIP	Ø.73ØE-Ø3	Ø.555E-Ø1
A68	PLATFORM SINKS	Ø.5ØØE-Ø3	Ø.456E-Ø1
A69	WASTE NOT RECOVERED	Ø.9ØØE ØØ	Ø.456E-Ø1
A19	SHIP SINKS DURING VOYAGE	Ø.5ØØE-Ø3	Ø.38ØE-Ø1
A57	DISCONNECTS PREMATURELY	Ø.15ØE-Ø2	Ø.3Ø4E-Ø1
A16	RECOVERY OF FLASKS IMPOSSIBLE	Ø.2ØØE ØØ	Ø.124E-Ø1
A15	FLASK DROPPED IN OPEN SEA	Ø.61ØE-Ø3	Ø.124E-Ø1
A58	HOIST SYSTEM RELEASES PIPE ASSEMBLY	Ø.2ØØE-Ø3	Ø.4Ø5E-Ø2
A13	SHIP SINKS NEAR SHORE	Ø.1ØØE-Ø2	Ø.2Ø3E-Ø2
A14	RECOVERY OF FLASKS IMPOSSIBLE	Ø.2ØØE-Ø1	Ø.2Ø3E-Ø2
A4Ø	STRINGER RECOVERY IMPOSSIBLE	Ø.2ØØE ØØ	Ø.122E-Ø2
A47	HOIST IS NOT ATTACHED	Ø.3ØØE ØØ	Ø.122E-Ø2
A49	MECHANICAL FAILURE	Ø.1ØØE-Ø3	Ø.6Ø8E-Ø3
A5Ø	HUMAN ERROR	Ø.1ØØE-Ø3	Ø.6Ø8E-Ø3
A3	FLASK BREACHED BY IMPACT	Ø.5ØØE-Ø5	Ø.5Ø7E-Ø3
A4	FLASK BREACHED BY FIRE	Ø.5ØØE-Ø5	Ø.5Ø7E-Ø3
A9	BOREHOLE NOT COMPLETED SATISFACTORILY	Ø.5ØØE-Ø5	Ø.5Ø7E-Ø3
A18	RECOVERY OF FLASK IMPOSSIBLE	Ø.2ØØE-Ø1	Ø.4Ø5E-Ø3
A17	FLASK DROPPED NEAR SHORE	Ø.2ØØE-Ø3	Ø.4Ø5E-Ø3
A84	RECOVERY IMPOSSIBLE BEFORE BEACHING	Ø.5ØØE-Ø3	Ø.3Ø5E-Ø3
A71	WASTE NOT RECOVERED	Ø.1ØØE-Ø1	Ø.3Ø5E-Ø3
A67	ROOF SHIELD NOT FULLY CLOSED	Ø.1ØØE-Ø2	Ø.253E-Ø3
A65	WATER LEVEL FALLS IN WET CELL	Ø.2ØØE-Ø2	Ø.2Ø3E-Ø3
A37	CANISTER RECOVERY IMPOSSIBLE	Ø.9ØØE ØØ	Ø.187E-Ø3
A35	CANISTER DROPS IN WET CELL	Ø.41ØE ØØ	Ø.187E-Ø3
A36	CANISTER SLIPS THRO' MOONPOOL	Ø.5ØØE-Ø5	Ø.187E-Ø3
A85	DP SYSTEM FAULTY	Ø.3ØØE ØØ	Ø.152E-Ø3
A62	FLASK OPENED INADVERTENTLY	Ø.1ØØE-Ø5	Ø.1Ø1E-Ø3
A1Ø	FLASK IN HARBOUR NOT RECOVERED	Ø.1ØØE-Ø5	Ø.1Ø1E-Ø3
A6	WASTE RELEASED FROM DAMAGED CONTAINER	Ø.1ØØE-Ø5	Ø.1Ø1E-Ø3
A86	INSUFFICIENT POWER FROM THRUSTERS	Ø.2ØØE ØØ	Ø.1Ø1E-Ø3
A21	FIRE SINKS SHIP	Ø.1ØØE-Ø5	Ø.76ØE-Ø4
A87	POWER LOST TO THRUSTERS	Ø.1ØØE ØØ	Ø.5Ø7E-Ø4
A66	CANISTER OR PIPE LIFTED TOO HIGH	Ø.5ØØE-Ø3	Ø.5Ø7E-Ø4
A91	FIRE/EXPLOSION	Ø.1ØØE-Ø2	Ø.5Ø7E-Ø6

DEEP DRILLED EMPLACEMENT
CASE 2 - NO RECOVERY FROM DEEP WATER

FAULT TREE

25 gates, 40 basic events.

```
A1                 RELEASE OF RADIOACTIVITY
     due to: A2        RADIOACTIVE FLASK LOST IN OCEAN
          OR A3        FLASK BREACHED BY IMPACT
          OR A4        FLASK BREACHED BY FIRE
          OR A5        CANISTERS LOST
          OR A6        WASTE RELEASED FROM DAMAGED CONTAINER
          OR A7        CANISTER EXPOSED
          OR A8        WASTE LOST WITH PLATFORM
          OR A9        BOREHOLE NOT COMPLETED SATISFACTORILY

A2                 RADIOACTIVE FLASK LOST IN OCEAN
     due to: A10       FLASK IN HARBOUR NOT RECOVERED
          OR A111      DUMMY
          OR A112      DUMMY
          OR A113      DUMMY
          OR A114      DUMMY

A111               DUMMY
     due to: A11       SUPPLY SHIP SINKS IN OPEN SEA
         AND A12       RECOVERY OF FLASKS IMPOSSIBLE

A11                SUPPLY SHIP SINKS IN OPEN SEA
     due to: A19       SHIP SINKS DURING VOYAGE
          OR A20       SHIP SINKS NEAR PLATFORM

A112               DUMMY
     due to: A13       SHIP SINKS NEAR SHORE
         AND A14       RECOVERY OF FLASKS IMPOSSIBLE

A113               DUMMY
     due to: A15       FLASK DROPPED IN OPEN SEA
         AND A16       RECOVERY OF FLASKS IMPOSSIBLE

A114               DUMMY
     due to: A17       FLASK DROPPED NEAR SHORE
         AND A18       RECOVERY OF FLASK IMPOSSIBLE

A20                SHIP SINKS NEAR PLATFORM
     due to: A21       FIRE SINKS SHIP
          OR A22       IMPACT SINKS SHIP

A5                 CANISTERS LOST
     due to: A115      DUMMY
          OR A116      DUMMY
          OR A117      DUMMY

A115               DUMMY
     due to: A35       CANISTER DROPS IN WET CELL
         AND A36       CANISTER SLIPS THRO' MOONPOOL
         AND A37       CANISTER RECOVERY IMPOSSIBLE
```

DEEP DRILLED EMPLACEMENT
CASE 2 - NO RECOVERY FROM DEEP WATER

A116 DUMMY
due to: A38 STRINGER PIPE DROPS THRO' MOONPOOL
AND A40 STRINGER RECOVERY IMPOSSIBLE

A38 STRINGER PIPE DROPS THRO' MOONPOOL
due to: A47 HOIST IS NOT ATTACHED
AND A48 HYDRAULIC JACK RELEASES STRINGER

A48 HYDRAULIC JACK RELEASES STRINGER
due to: A49 MECHANICAL FAILURE
OR A50 HUMAN ERROR

A117 DUMMY
due to: A41 STRINGER PIPE RELEASED DURING LOWERING
AND A42 STRINGER RECOVERY IMPOSSIBLE

A41 STRINGER PIPE RELEASED DURING LOWERING
due to: A58 HOIST SYSTEM RELEASES PIPE ASSEMBLY
OR A57 DISCONNECTS PREMATURELY
OR A51 PIPE ASSEMBLY BREAKS

A51 PIPE ASSEMBLY BREAKS
due to: A52 PIPE FAILS
OR A118 DUMMY

A118 DUMMY
due to: A54 PLATFORM DRIFTS WITH STRING IN HOLE
AND A56 EMERGENCY OPERATION INEFFECTIVE

A7 CANISTER EXPOSED
due to: A62 FLASK OPENED INADVERTENTLY
OR A120 DUMMY

A120 DUMMY
due to: A67 ROOF SHIELD NOT FULLY CLOSED
AND A121 DUMMY

A121 DUMMY
due to: A65 WATER LEVEL FALLS IN WET CELL
OR A66 CANISTER OR PIPE LIFTED TOO HIGH

A8 WASTE LOST WITH PLATFORM
due to: A122 DUMMY
OR A123 DUMMY

A122 DUMMY
due to: A68 PLATFORM SINKS
AND A69 WASTE NOT RECOVERED

A123 DUMMY
due to: A70 PLATFORM IS STRANDED
AND A71 WASTE NOT RECOVERED

DEEP DRILLED EMPLACEMENT
CASE 2 - NO RECOVERY FROM DEEP WATER

A7Ø PLATFORM IS STRANDED

due to: A83 PLATFORM LOSES STATION
AND A84 RECOVERY IMPOSSIBLE BEFORE BEACHING

A83 PLATFORM LOSES STATION

due to: A85 DP SYSTEM FAULTY
OR A86 INSUFFICIENT POWER FROM THRUSTERS
OR A87 POWER LOST TO THRUSTERS
OR A91 FIRE/EXPLOSION

DEEP DRILLED EMPLACEMENT
CASE 2 - NO RECOVERY FROM DEEP WATER

	BASIC EVENTS	FAILURE RATE
A10	FLASK IN HARBOUR NOT RECOVERED	0.100E-05
A12	RECOVERY OF FLASKS IMPOSSIBLE	0.999E 00
A13	SHIP SINKS NEAR SHORE	0.100E-02
A14	RECOVERY OF FLASKS IMPOSSIBLE	0.200E-01
A15	FLASK DROPPED IN OPEN SEA	0.610E-03
A16	RECOVERY OF FLASKS IMPOSSIBLE	0.999E 00
A17	FLASK DROPPED NEAR SHORE	0.200E-03
A18	RECOVERY OF FLASK IMPOSSIBLE	0.200E-01
A19	SHIP SINKS DURING VOYAGE	0.500E-03
A21	FIRE SINKS SHIP	0.100E-05
A22	IMPACT SINKS SHIP	0.730E-03
A3	FLASK BREACHED BY IMPACT	0.500E-05
A4	FLASK BREACHED BY FIRE	0.500E-05
A35	CANISTER DROPS IN WET CELL	0.410E 00
A36	CANISTER SLIPS THRO' MOONPOOL	0.500E-05
A37	CANISTER RECOVERY IMPOSSIBLE	0.999E 00
A40	STRINGER RECOVERY IMPOSSIBLE	0.999E 00
A47	HOIST IS NOT ATTACHED	0.300E 00
A49	MECHANICAL FAILURE	0.100E-03
A50	HUMAN ERROR	0.100E-03
A42	STRINGER RECOVERY IMPOSSIBLE	0.999E 00
A58	HOIST SYSTEM RELEASES PIPE ASSEMBLY	0.200E-03
A57	DISCONNECTS PREMATURELY	0.150E-02
A52	PIPE FAILS	0.300E-01
A54	PLATFORM DRIFTS WITH STRING IN HOLE	0.500E-01
A56	EMERGENCY OPERATION INEFFECTIVE	0.200E 00
A6	WASTE RELEASED FROM DAMAGED CONTAINER	0.100E-05
A62	FLASK OPENED INADVERTENTLY	0.100E-05
A67	ROOF SHIELD NOT FULLY CLOSED	0.100E-02
A65	WATER LEVEL FALLS IN WET CELL	0.200E-02
A66	CANISTER OR PIPE LIFTED TOO HIGH	0.500E-03
A68	PLATFORM SINKS	0.500E-03
A69	WASTE NOT RECOVERED	0.999E 00
A71	WASTE NOT RECOVERED	0.100E-01
A84	RECOVERY IMPOSSIBLE BEFORE BEACHING	0.500E-03
A85	DP SYSTEM FAULTY	0.300E 00
A86	INSUFFICIENT POWER FROM THRUSTERS	0.200E 00
A87	POWER LOST TO THRUSTERS	0.100E 00
A91	FIRE/EXPLOSION	0.100E-02
A9	BOREHOLE NOT COMPLETED SATISFACTORILY	0.500E-05

Probability limit = 0.000E 00 Maximum length of cut set = 10

DEEP DRILLED EMPLACEMENT
CASE 2 - NO RECOVERY FROM DEEP WATER

Cut set listing for A1 : RELEASE OF RADIOACTIVITY
26 minimal cut sets

Probability	Cut set list		
Ø.3ØØE-Ø1	A42	A52	
Ø.999E-Ø2	A42	A54	A56
Ø.15ØE-Ø2	A42	A57	
Ø.729E-Ø3	A12	A22	
Ø.6Ø9E-Ø3	A15	A16	
Ø.499E-Ø3	A12	A19	
Ø.499E-Ø3	A68	A69	
Ø.2ØØE-Ø3	A42	A58	
Ø.3ØØE-Ø4	A4Ø	A47	A5Ø
Ø.3ØØE-Ø4	A4Ø	A47	A49
Ø.2ØØE-Ø4	A13	A14	
Ø.5ØØE-Ø5	A4		
Ø.5ØØE-Ø5	A3		
Ø.5ØØE-Ø5	A9		
Ø.4ØØE-Ø5	A17	A18	
Ø.2Ø5E-Ø5	A35	A36	A37
Ø.2ØØE-Ø5	A67	A65	
Ø.15ØE-Ø5	A71	A84	A85
Ø.1ØØE-Ø5	A6		
Ø.1ØØE-Ø5	A62		
Ø.1ØØE-Ø5	A1Ø		
Ø.1ØØE-Ø5	A71	A84	A86
Ø.999E-Ø6	A12	A21	
Ø.5ØØE-Ø6	A71	A84	A87
Ø.5ØØE-Ø6	A67	A66	
Ø.5ØØE-Ø8	A71	A84	A91

OVERALL PROBABILITY OF FAILURE = Ø.436E-Ø1

DEEP DRILLED EMPLACEMENT
CASE 2 - NO RECOVERY FROM DEEP WATER

ANALYSIS OF IMPORTANCE

	Basic Event	Failure Rate	Importance
A42	STRINGER RECOVERY IMPOSSIBLE	Ø.999E ØØ	Ø.946E ØØ
A52	PIPE FAILS	Ø.3ØØE-Ø1	Ø.687E ØØ
A54	PLATFORM DRIFTS WITH STRING IN HOLE	Ø.5ØØE-Ø1	Ø.229E ØØ
A56	EMERGENCY OPERATION INEFFECTIVE	Ø.2ØØE ØØ	Ø.229E ØØ
A57	DISCONNECTS PREMATURELY	Ø.15ØE-Ø2	Ø.343E-Ø1
A12	RECOVERY OF FLASKS IMPOSSIBLE	Ø.999E ØØ	Ø.282E-Ø1
A22	IMPACT SINKS SHIP	Ø.73ØE-Ø3	Ø.167E-Ø1
A15	FLASK DROPPED IN OPEN SEA	Ø.61ØE-Ø3	Ø.14ØE-Ø1
A16	RECOVERY OF FLASKS IMPOSSIBLE	Ø.999E ØØ	Ø.14ØE-Ø1
A19	SHIP SINKS DURING VOYAGE	Ø.5ØØE-Ø3	Ø.114E-Ø1
A68	PLATFORM SINKS	Ø.5ØØE-Ø3	Ø.114E-Ø1
A69	WASTE NOT RECOVERED	Ø.999E ØØ	Ø.114E-Ø1
A58	HOIST SYSTEM RELEASES PIPE ASSEMBLY	Ø.2ØØE-Ø3	Ø.458E-Ø2
A4Ø	STRINGER RECOVERY IMPOSSIBLE	Ø.999E ØØ	Ø.137E-Ø2
A47	HOIST IS NOT ATTACHED	Ø.3ØØE ØØ	Ø.137E-Ø2
A49	MECHANICAL FAILURE	Ø.1ØØE-Ø3	Ø.687E-Ø3
A5Ø	HUMAN ERROR	Ø.1ØØE-Ø3	Ø.687E-Ø3
A13	SHIP SINKS NEAR SHORE	Ø.1ØØE-Ø2	Ø.458E-Ø3
A14	RECOVERY OF FLASKS IMPOSSIBLE	Ø.2ØØE-Ø1	Ø.458E-Ø3
A3	FLASK BREACHED BY IMPACT	Ø.5ØØE-Ø5	Ø.115E-Ø3
A4	FLASK BREACHED BY FIRE	Ø.5ØØE-Ø5	Ø.115E-Ø3
A9	BOREHOLE NOT COMPLETED SATISFACTORILY	Ø.5ØØE-Ø5	Ø.115E-Ø3
A18	RECOVERY OF FLASK IMPOSSIBLE	Ø.2ØØE-Ø1	Ø.917E-Ø4
A17	FLASK DROPPED NEAR SHORE	Ø.2ØØE-Ø3	Ø.917E-Ø4
A84	RECOVERY IMPOSSIBLE BEFORE BEACHING	Ø.5ØØE-Ø3	Ø.689E-Ø4
A71	WASTE NOT RECOVERED	Ø.1ØØE-Ø1	Ø.689E-Ø4
A67	ROOF SHIELD NOT FULLY CLOSED	Ø.1ØØE-Ø2	Ø.573E-Ø4
A36	CANISTER SLIPS THRO' MOONPOOL	Ø.5ØØE-Ø5	Ø.469E-Ø4
A37	CANISTER RECOVERY IMPOSSIBLE	Ø.999E ØØ	Ø.469E-Ø4
A35	CANISTER DROPS IN WET CELL	Ø.41ØE ØØ	Ø.469E-Ø4
A65	WATER LEVEL FALLS IN WET CELL	Ø.2ØØE-Ø2	Ø.458E-Ø4
A85	DP SYSTEM FAULTY	Ø.3ØØE ØØ	Ø.344E-Ø4
A62	FLASK OPENED INADVERTENTLY	Ø.1ØØE-Ø5	Ø.229E-Ø4
A1Ø	FLASK IN HARBOUR NOT RECOVERED	Ø.1ØØE-Ø5	Ø.229E-Ø4
A6	WASTE RELEASED FROM DAMAGED CONTAINER	Ø.1ØØE-Ø5	Ø.229E-Ø4
A86	INSUFFICIENT POWER FROM THRUSTERS	Ø.2ØØE ØØ	Ø.229E-Ø4
A21	FIRE SINKS SHIP	Ø.1ØØE-Ø5	Ø.229E-Ø4
A87	POWER LOST TO THRUSTERS	Ø.1ØØE ØØ	Ø.115E-Ø4
A66	CANISTER OR PIPE LIFTED TOO HIGH	Ø.5ØØE-Ø3	Ø.115E-Ø4
A91	FIRE/EXPLOSION	Ø.1ØØE-Ø2	Ø.115E-Ø6

DEEP DRILLED EMPLACEMENT
CASE 3 - NO RECOVERY FROM DEEP WATER, MAGNITUDE INCLUDED

FAULT TREE

25 gates, 40 basic events.

```
A1                      RELEASE OF RADIOACTIVITY
     due to: A2            RADIOACTIVE FLASK LOST IN OCEAN
          OR A3            FLASK BREACHED BY IMPACT
          OR A4            FLASK BREACHED BY FIRE
          OR A5            CANISTERS LOST
          OR A6            WASTE RELEASED FROM DAMAGED CONTAINER
          OR A7            CANISTER EXPOSED
          OR A8            WASTE LOST WITH PLATFORM
          OR A9            BOREHOLE NOT COMPLETED SATISFACTORILY

A2                      RADIOACTIVE FLASK LOST IN OCEAN
     due to: A10           FLASK IN HARBOUR NOT RECOVERED
          OR A111          DUMMY
          OR A112          DUMMY
          OR A113          DUMMY
          OR A114          DUMMY

A111                    DUMMY
     due to: A11           SUPPLY SHIP SINKS IN OPEN SEA
         AND A12           RECOVERY OF FLASKS IMPOSSIBLE

A11                     SUPPLY SHIP SINKS IN OPEN SEA
     due to: A19           SHIP SINKS DURING VOYAGE
          OR A20           SHIP SINKS NEAR PLATFORM

A112                    DUMMY
     due to: A13           SHIP SINKS NEAR SHORE
         AND A14           RECOVERY OF FLASKS IMPOSSIBLE

A113                    DUMMY
     due to: A15           FLASK DROPPED IN OPEN SEA
         AND A16           RECOVERY OF FLASKS IMPOSSIBLE

A114                    DUMMY
     due to: A17           FLASK DROPPED NEAR SHORE
         AND A18           RECOVERY OF FLASK IMPOSSIBLE

A20                     SHIP SINKS NEAR PLATFORM
     due to: A21           FIRE SINKS SHIP
          OR A22           IMPACT SINKS SHIP

A5                      CANISTERS LOST
     due to: A115          DUMMY
          OR A116          DUMMY
          OR A117          DUMMY

A115                    DUMMY
     due to: A35           CANISTER DROPS IN WET CELL
         AND A36           CANISTER SLIPS THRO' MOONPOOL
         AND A37           CANISTER RECOVERY IMPOSSIBLE
```

DEEP DRILLED EMPLACEMENT

CASE 3 - NO RECOVERY FROM DEEP WATER, MAGNITUDE INCLUDED

```
A116                        DUMMY
      due to: A38               STRINGER PIPE DROPS THRO' MOONPOOL
          AND A4Ø               STRINGER RECOVERY IMPOSSIBLE

A38                         STRINGER PIPE DROPS THRO' MOONPOOL
      due to: A47               HOIST IS NOT ATTACHED
          AND A48               HYDRAULIC JACK RELEASES STRINGER

A48                         HYDRAULIC JACK RELEASES STRINGER
      due to: A49               MECHANICAL FAILURE
          OR  A5Ø               HUMAN ERROR

A117                        DUMMY
      due to: A41               STRINGER PIPE RELEASED DURING LOWERING
          AND A42               STRINGER RECOVERY IMPOSSIBLE

A41                         STRINGER PIPE RELEASED DURING LOWERING
      due to: A58               HOIST SYSTEM RELEASES PIPE ASSEMBLY
          OR  A57               DISCONNECTS PREMATURELY
          OR  A51               PIPE ASSEMBLY BREAKS

A51                         PIPE ASSEMBLY BREAKS
      due to: A52               PIPE FAILS
          OR  A118              DUMMY

A118                        DUMMY
      due to: A54               PLATFORM DRIFTS WITH STRING IN HOLE
          AND A56               EMERGENCY OPERATION INEFFECTIVE

A7                          CANISTER EXPOSED
      due to: A62               FLASK OPENED INADVERTENTLY
          OR  A12Ø              DUMMY

A12Ø                        DUMMY
      due to: A67               ROOF SHIELD NOT FULLY CLOSED
          AND A121              DUMMY

A121                        DUMMY
      due to: A65               WATER LEVEL FALLS IN WET CELL
          OR  A66               CANISTER OR PIPE LIFTED TOO HIGH

A8                          WASTE LOST WITH PLATFORM
      due to: A122              DUMMY
          OR  A123              DUMMY

A122                        DUMMY
      due to: A68               PLATFORM SINKS
          AND A69               WASTE NOT RECOVERED

A123                        DUMMY
      due to: A7Ø               PLATFORM IS STRANDED
          AND A71               WASTE NOT RECOVERED
```

DEEP DRILLED EMPLACEMENT

CASE 3 - NO RECOVERY FROM DEEP WATER, MAGNITUDE INCLUDED

A7Ø		PLATFORM IS STRANDED
	due to: A83	PLATFORM LOSES STATION
	AND A84	RECOVERY IMPOSSIBLE BEFORE BEACHING
A83		PLATFORM LOSES STATION
	due to: A85	DP SYSTEM FAULTY
	OR A86	INSUFFICIENT POWER FROM THRUSTERS
	OR A87	POWER LOST TO THRUSTERS
	OR A91	FIRE/EXPLOSION

DEEP DRILLED EMPLACEMENT

CASE 3 - NO RECOVERY FROM DEEP WATER, MAGNITUDE INCLUDED

	BASIC EVENTS	FAILURE RATE
A10	FLASK IN HARBOUR NOT RECOVERED	0.100E-07
A12	RECOVERY OF FLASKS IMPOSSIBLE	0.999E 00
A13	SHIP SINKS NEAR SHORE	0.100E-03
A14	RECOVERY OF FLASKS IMPOSSIBLE	0.200E-01
A15	FLASK DROPPED IN OPEN SEA	0.610E-05
A16	RECOVERY OF FLASKS IMPOSSIBLE	0.999E 00
A17	FLASK DROPPED NEAR SHORE	0.200E-05
A18	RECOVERY OF FLASK IMPOSSIBLE	0.200E-01
A19	SHIP SINKS DURING VOYAGE	0.500E-04
A21	FIRE SINKS SHIP	0.100E-06
A22	IMPACT SINKS SHIP	0.730E-04
A3	FLASK BREACHED BY IMPACT	0.500E-07
A4	FLASK BREACHED BY FIRE	0.500E-06
A35	CANISTER DROPS IN WET CELL	0.160E-03
A36	CANISTER SLIPS THRO' MOONPOOL	0.500E-05
A37	CANISTER RECOVERY IMPOSSIBLE	0.999E 00
A40	STRINGER RECOVERY IMPOSSIBLE	0.999E 00
A47	HOIST IS NOT ATTACHED	0.300E 00
A49	MECHANICAL FAILURE	0.200E-05
A50	HUMAN ERROR	0.200E-05
A42	STRINGER RECOVERY IMPOSSIBLE	0.999E 00
A58	HOIST SYSTEM RELEASES PIPE ASSEMBLY	0.400E-05
A57	DISCONNECTS PREMATURELY	0.300E-04
A52	PIPE FAILS	0.600E-03
A54	PLATFORM DRIFTS WITH STRING IN HOLE	0.100E-02
A56	EMERGENCY OPERATION INEFFECTIVE	0.200E 00
A6	WASTE RELEASED FROM DAMAGED CONTAINER	0.400E-09
A62	FLASK OPENED INADVERTENTLY	0.400E-09
A67	ROOF SHIELD NOT FULLY CLOSED	0.100E-02
A65	WATER LEVEL FALLS IN WET CELL	0.800E-05
A66	CANISTER OR PIPE LIFTED TOO HIGH	0.200E-05
A68	PLATFORM SINKS	0.500E-04
A69	WASTE NOT RECOVERED	0.999E 00
A71	WASTE NOT RECOVERED	0.100E-01
A84	RECOVERY IMPOSSIBLE BEFORE BEACHING	0.500E-03
A85	DP SYSTEM FAULTY	0.300E-01
A86	INSUFFICIENT POWER FROM THRUSTERS	0.200E-01
A87	POWER LOST TO THRUSTERS	0.100E-01
A91	FIRE/EXPLOSION	0.100E-03
A9	BOREHOLE NOT COMPLETED SATISFACTORILY	0.500E-06

Probability limit = 0.000E 00   Maximum length of cut set = 10

DEEP DRILLED EMPLACEMENT
CASE 3 - NO RECOVERY FROM DEEP WATER, MAGNITUDE INCLUDED

Cut set listing for A1 : RELEASE OF RADIOACTIVITY
26 minimal cut sets

Probability	Cut set list		
Ø.599E-Ø3	A42	A52	
Ø.2ØØE-Ø3	A42	A54	A56
Ø.729E-Ø4	A12	A22	
Ø.499E-Ø4	A12	A19	
Ø.499E-Ø4	A68	A69	
Ø.3ØØE-Ø4	A42	A57	
Ø.6Ø9E-Ø5	A15	A16	
Ø.4ØØE-Ø5	A42	A58	
Ø.2ØØE-Ø5	A13	A14	
Ø.599E-Ø6	A4Ø	A47	A5Ø
Ø.599E-Ø6	A4Ø	A47	A49
Ø.5ØØE-Ø6	A4		
Ø.5ØØE-Ø6	A9		
Ø.15ØE-Ø6	A71	A84	A85
Ø.1ØØE-Ø6	A71	A84	A86
Ø.999E-Ø7	A12	A21	
Ø.5ØØE-Ø7	A3		
Ø.5ØØE-Ø7	A71	A84	A87
Ø.4ØØE-Ø7	A17	A18	
Ø.1ØØE-Ø7	A1Ø		
Ø.8ØØE-Ø8	A67	A65	
Ø.2ØØE-Ø8	A67	A66	
Ø.799E-Ø9	A35	A36	A37
Ø.5ØØE-Ø9	A71	A84	A91
Ø.4ØØE-Ø9	A6		
Ø.4ØØE-Ø9	A62		

OVERALL PROBABILITY OF FAILURE = Ø.1Ø2E-Ø2
= 2.75 Canisters/year

DEEP DRILLED EMPLACEMENT

CASE 3 - NO RECOVERY FROM DEEP WATER, MAGNITUDE INCLUDED

ANALYSIS OF IMPORTANCE

	Basic Event	Failure Rate	Importance
A42	STRINGER RECOVERY IMPOSSIBLE	0.999E 00	0.820E 00
A52	PIPE FAILS	0.600E-03	0.590E 00
A54	PLATFORM DRIFTS WITH STRING IN HOLE	0.100E-02	0.197E 00
A56	EMERGENCY OPERATION INEFFECTIVE	0.200E 00	0.197E 00
A12	RECOVERY OF FLASKS IMPOSSIBLE	0.999E 00	0.121E 00
A22	IMPACT SINKS SHIP	0.730E-04	0.717E-01
A19	SHIP SINKS DURING VOYAGE	0.500E-04	0.491E-01
A68	PLATFORM SINKS	0.500E-04	0.491E-01
A69	WASTE NOT RECOVERED	0.999E 00	0.491E-01
A57	DISCONNECTS PREMATURELY	0.300E-04	0.295E-01
A16	RECOVERY OF FLASKS IMPOSSIBLE	0.999E 00	0.600E-02
A15	FLASK DROPPED IN OPEN SEA	0.610E-05	0.600E-02
A58	HOIST SYSTEM RELEASES PIPE ASSEMBLY	0.400E-05	0.393E-02
A13	SHIP SINKS NEAR SHORE	0.100E-03	0.197E-02
A14	RECOVERY OF FLASKS IMPOSSIBLE	0.200E-01	0.197E-02
A40	STRINGER RECOVERY IMPOSSIBLE	0.999E 00	0.118E-02
A47	HOIST IS NOT ATTACHED	0.300E 00	0.118E-02
A49	MECHANICAL FAILURE	0.200E-05	0.590E-03
A50	HUMAN ERROR	0.200E-05	0.590E-03
A4	FLASK BREACHED BY FIRE	0.500E-06	0.492E-03
A9	BOREHOLE NOT COMPLETED SATISFACTORILY	0.500E-06	0.492E-03
A84	RECOVERY IMPOSSIBLE BEFORE BEACHING	0.500E-03	0.296E-03
A71	WASTE NOT RECOVERED	0.100E-01	0.296E-03
A85	DP SYSTEM FAULTY	0.300E-01	0.148E-03
A86	INSUFFICIENT POWER FROM THRUSTERS	0.200E-01	0.984E-04
A21	FIRE SINKS SHIP	0.100E-06	0.983E-04
A3	FLASK BREACHED BY IMPACT	0.500E-07	0.492E-04
A87	POWER LOST TO THRUSTERS	0.100E-01	0.492E-04
A18	RECOVERY OF FLASK IMPOSSIBLE	0.200E-01	0.394E-04
A17	FLASK DROPPED NEAR SHORE	0.200E-05	0.394E-04
A10	FLASK IN HARBOUR NOT RECOVERED	0.100E-07	0.984E-05
A67	ROOF SHIELD NOT FULLY CLOSED	0.100E-02	0.984E-05
A65	WATER LEVEL FALLS IN WET CELL	0.800E-05	0.787E-05
A66	CANISTER OR PIPE LIFTED TOO HIGH	0.200E-05	0.197E-05
A35	CANISTER DROPS IN WET CELL	0.160E-03	0.786E-06
A36	CANISTER SLIPS THRO' MOONPOOL	0.500E-05	0.786E-06
A37	CANISTER RECOVERY IMPOSSIBLE	0.999E 00	0.786E-06
A91	FIRE/EXPLOSION	0.100E-03	0.492E-06
A6	WASTE RELEASED FROM DAMAGED CONTAINER	0.400E-09	0.394E-06
A62	FLASK OPENED INADVERTENTLY	0.400E-09	0.394E-06

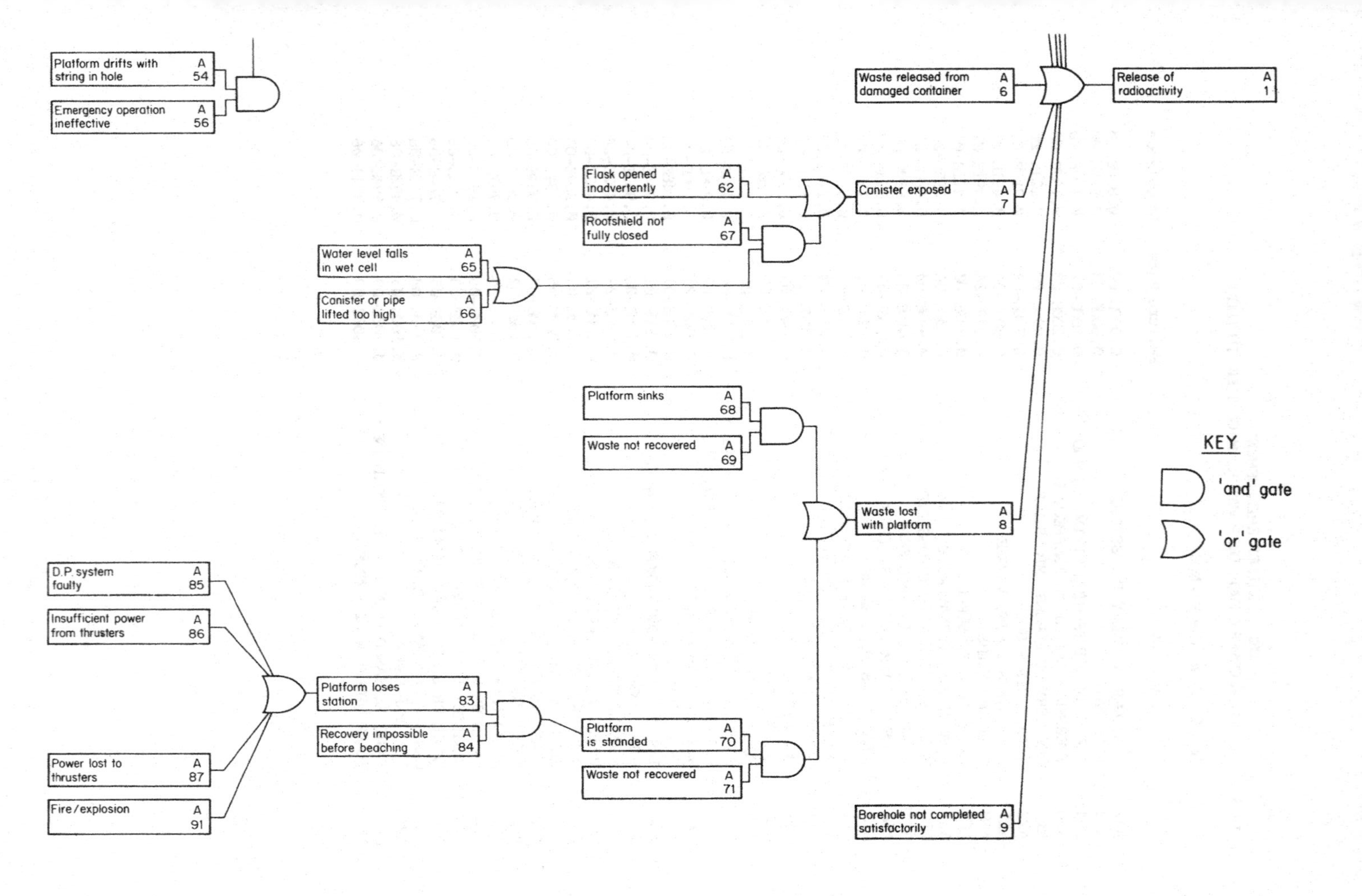
Platform drifts with string in hole A 54
Emergency operation ineffective A 56
Waste released from damaged container A 6
Release of radioactivity A 1
Flask opened inadvertently A 62
Roofshield not fully closed A 67
Canister exposed A 7
Water level falls in wet cell A 65
Canister or pipe lifted too high A 66
Platform sinks A 68
Waste not recovered A 69
Waste lost with platform A 8
KEY
'and' gate
'or' gate
D.P. system faulty A 85
Insufficient power from thrusters A 86
Power lost to thrusters A 87
Fire/explosion A 91
Platform loses station A 83
Recovery impossible before beaching A 84
Platform is stranded A 70
Waste not recovered A 71
Borehole not completed satisfactorily A 9

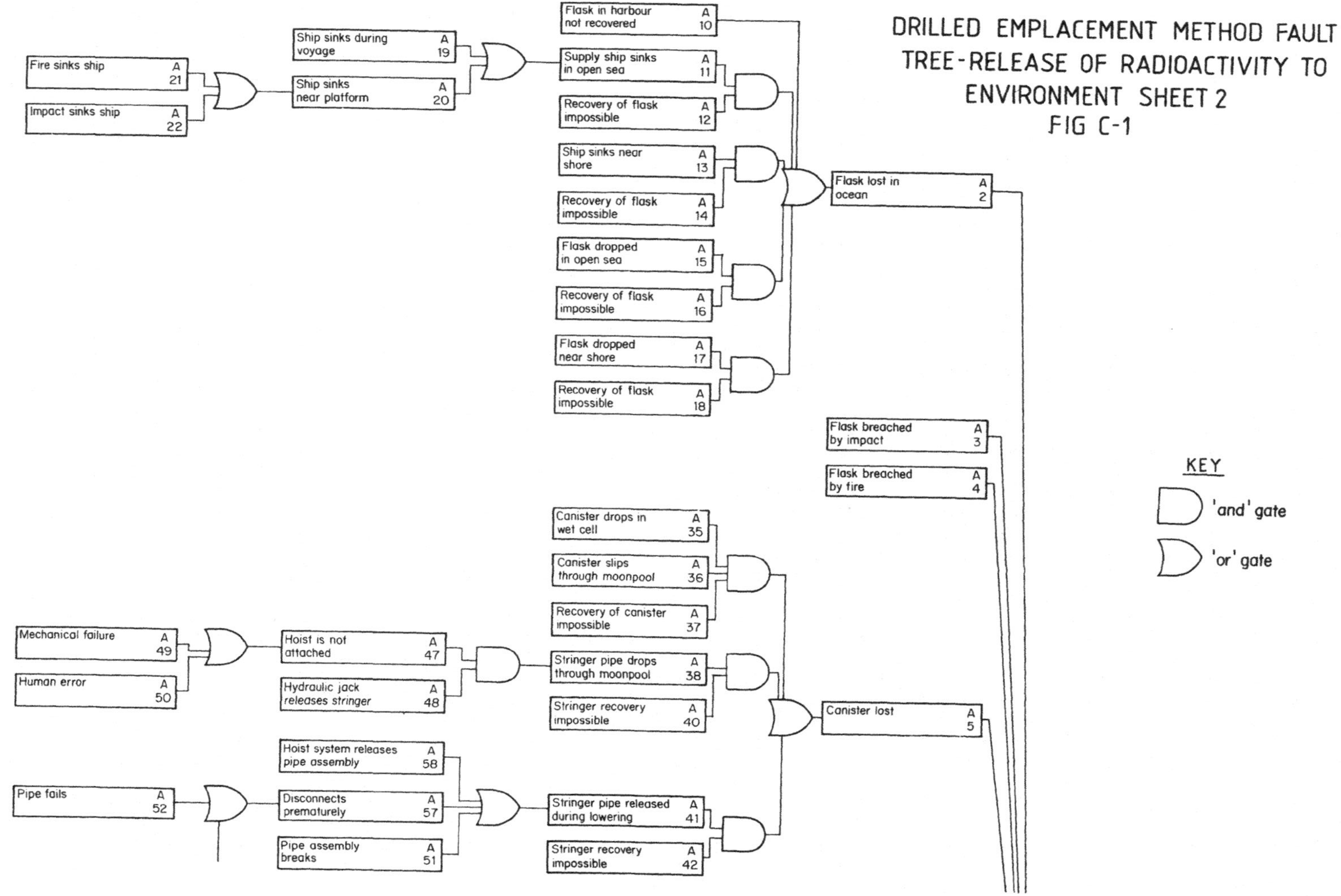

DRILLED EMPLACEMENT METHOD FAULT TREE-RELEASE OF RADIOACTIVITY TO ENVIRONMENT SHEET 2
FIG C-1

APPENDIX D

FAULT TREE ANALYSIS

HAZARD TO PERSONNEL

SUMMARY OF COMPUTER OUTPUT
(T.W PROGRAM "CUTS")

DRILLED EMPLACEMENT METHOD

CONTENTS

APPENDIX D DRILLED EMPLACEMENT METHOD

FAULT TREE ANALYSIS - HAZARD TO PERSONNEL

DATA ANALYSIS - EXPLANATORY NOTES

LOGIC REFERENCE	DERIVATION OF FAILURE RATES OR PROBABILITY	ASSESSMENT (θ = f/year) (p = f probability)
B.1	From Figure 1. logic. This assessment includes all hazards to personnel when on platform that lead to death in a short time, i.e. operational and radiation hazards. The assessment also includes the hazards to platform personnel while at sea on a passengership (being ferried to or from the platform), which also lead to "Short time Death".	θ = 6.2 E - 1
B.120	From B7, B8 and B9.	θ = 8.43 E - 5
B.121	It is assumed for these studies that 5 people may suffer short time death in an accident of exposure to critical radiation. However it is a pessimistic figure and at present can not be seen to occur with the designed safety standards applied on platform.	θ = 5
B.2	From B120 and B121. The assessment of life lost due to critical radiation dose uptake forms the major work of this study. This θ value is calculated in the following way: The probability of an accident of critical radiation uptake from fig.120 is p = 8.43 E - 5. The total number of people, out of 500 platform personnel who are directly involved	0 = 4.2 E - 4

<table>
<tr><th>LOGIC REFERENCE</th><th>DERIVATION OF FAILURE RATES OR PROBABILITY</th><th>ASSESSMENT
(θ = f/year)
(p = f probability)</th></tr>
<tr><td>B.2 Cont.</td><td>with the HLW handling, inspection and maintenance, where such an accident can happen, is not more than 100. It is assumed however that during any accident leading to critical radiation uptake not more than 5 persons (extremely pessimistic figure) would suffer short time death. Thus $\theta = p \times 5 = 8.43\ E - 5 \times 5 = 4.2\ E - 4$.

The death risk p.a. for all 500 platform personnel (this number is used consistently in this study): $r = \theta \div 500 = 8.44\ E - 7$ per year (or 1 in 119000C per year).

If, on the other hand, the death risk per annum were related only to the 100 personnel directly involved in the high radiation area activities, the following value would result:

$$r = \frac{\theta}{100} = 4.2\ E - 6\ y^{-1}$$ (or 1 in 238000)

Even this risk is smaller by at least one order of magnitude than any of all the other risks involving no radioactivity (i.e. logic 3, 4, 5 & 6) and so forms an insignificant portion of the total risk.</td><td></td></tr>
</table>

LOGIC REFERENCE	DERIVATION OF FAILURE RATES OR PROBABILITY	ASSESSMENT (θ = f/year) (p = f probability)
B.3	The disposal platform will be similar to drilling and production platforms regarding operational activities at large. Therefore the actual death statistics for the British sector of the North Sea Offshore operations were used. The Department of Energy submits a report to Parliament annually entitled: "DEVELOPMENT OF OIL & GAS RESOURCES OF THE UK". Appendix 15 of the 1985 report lists in brief the serious and fatal accidents related to offshore installations and the number of people employed on them. For calculating θ, data given for the 12 year period between 1973 and 1984 were used as long term average. $\theta = 500 : \frac{181560}{112} = 0.31$ where 112 = number of dead persons. 181560 = total number of people employed. 500 = total number of personnel for platform to be employed. Risk of death: 6.2 E - 4 = 1 in 1612 per year. However, analysing the data in a more detailed way, the following should be noted:	θ = 3.1 E - 1

APPENDIX D DRILLED EMPLACEMENT METHOD

FAULT TREE ANALYSIS - HAZARD TO PERSONNEL

DATA ANALYSIS - EXPLANATORY NOTES

LOGIC REFERENCE	DERIVATION OF FAILURE RATES OR PROBABILITY	ASSESSMENT (θ = f/year) (p = f probability)
B.3 Cont.	a) The estimated number of personnel employed on the installation from 1980 is not directly comparable with the numbers shown for previous years, since from 1980 the number of people employed include construction workers, personnel of mobile drilling rigs, service vessels, support barges and survey teams. The appendix does not indicate clearly which group were included in the numbers shown for the period prior to 1980. b) The statistical data shown for the period prior to 1980 would give the following θ values: $\theta_1 = 500: \frac{57060}{53} = 4.6$ E - 1 (excluding death on vessels). $\theta_2 = 500: \frac{57060}{67} = 5.8$ E - 1 (excluding death on vessels). c) The statistical data shown for the period from 1980 would give a θ value of: $\theta_3 = 500: \frac{124500}{45} = 1.8$ E - 1 (includes the deaths on vessels and deaths on all mentioned installations).	

LOGIC REFERENCE	DERIVATION OF FAILURE RATES OR PROBABILITY	ASSESSMENT (θ = f/year) (p = f probability)
B.3 Cont.	d) Averaging $\theta_1 + \theta_3$ gives a value of 3.2 E - 1 which is the same as the long term average. e) The average of θ_2 and θ_3 gives a value of 3.8 E - 1 (The difference in order of magnitude is insignificant).	
B.4	The assessment value is calculated from the following: a) The probability of the platform sinking is taken from Appendix B, Logic Ref. 68 as 5 E - 4. b) It is assumed that at least 100 people will lose their lives (out of 250 on the platform) when platform sinks. (5 E - 4 x 100 = 5 E - 2)	θ = 5 E - 2
B.5	The ship will be dedicated primarily to ferry the platform personnel from shore to disposal platform and from platform back to shore. A round trip will take about 10 - 14 days. Working out the operational logic, it is calculated that this vessel will be at sea only for ½ year in each year and this fact is taken into account in the "Hazard Assessment".	θ = 2.5 E - 2

APPENDIX D DRILLED EMPLACEMENT METHOD

FAULT TREE ANALYSIS - HAZARD TO PERSONNEL

DATA ANALYSIS - EXPLANATORY NOTES

LOGIC REFERENCE	DERIVATION OF FAILURE RATES OR PROBABILITY	ASSESSMENT (θ = f/year) (p = f probability)
B.5 Cont.	The assessment value is arrived at as follows: a) The probability of ship sinking during voyage is taken from Appendix B, Logic Ref.19. as 5 E - 4. b) By operational logic, this ship is at sea only half the operating time. c) It is assumed that at least 100 people will lose their lives (out of 250 on board) when ship sinks. (5 E - 4 x 0.5 x 100 = 2.5 E - 2) This value is very much in agreement with the one calculated from the actual data for 1975 - 1980, given in the Department of Trade & Transport report, Table D-1. Number of deaths due to founderings is 37. Total number of people involved is 554,000. This would give a θ value of 50: $\frac{554000}{37}$ = 3.3 E - 2. (Thus the risk of death per year is r = 0 : 500 = 6.7 x E - 5 or 1 in 14973)	
B.6	The Dept. of Trade & Transport has a combined publication entitled: "Casualties to Vessels and Accidents to Man". Table D1 shows a summary of fatalities for period 1975 - 1980 collected from the above report. The data published are related to Seamen. No data is available for ship passenger accidents and deaths. The platform personnel on route to and from the platform are considered neither as passengers, nor	θ = 2.3 E - 1

LOGIC REFERENCE	DERIVATION OF FAILURE RATES OR PROBABILITY	ASSESSMENT (θ = f/year) (p = f probability)
B.6 Cont.	as Seamen in the true meaning of these words; they would fall into a definition between the two categories. Since data for passengers are not available, the statistics for seamen are analysed and related to the total number of platform personnel. Accordingly the θ value was calculated from: Number of Seamen involved 554000 Number of deaths 255 Number of platform personnel 500 $\theta = 500: \frac{554000}{255} = 2.3\ E-1$ These statistics do not include: 1. Number of deaths due to founderings and 2. Number of deaths in harbour.	

APPENDIX D DRILLED EMPLACEMENT METHOD

FAULT TREE ANALYSIS - HAZARD TO PERSONNEL

DATA ANALYSIS - EXPLANATORY NOTES

LOGIC REFERENCE	DERIVATION OF FAILURE RATES OR PROBABILITY	ASSESSMENT (θ = f/year) (p = f probability)
B.6 Cont.	Using the official statistics of 'seamen deaths' may seem somewhat pessimistic but since the deaths in harbour, which includes hazards during boarding and disembarkation, are not included (and these do happen to a good number of crew and passengers according to the Department of Transport) the θ value calculated is as good as any deducted by probability calculations.	
B.7	The only case considered in this logic is when the flask on deck opened inadvertantly. It is considered an extremely unlikely event and θ = E - 6 is taken from Appendix B, Logic Ref. 62. A suicide attempt may occur amongst the platform personnel but one man alone could not remove the flask lid to expose the canisters; furthermore tools necessary for such an operation are only available in the wet cell. So a collusion of a number of people, who are simultaneously in a deranged or disturbed state of mind, must occur first. All other catastrophic accidents like fire, sinking of platform etc., which would result in critical radiation exposure, are not considered for two major reasons: 1. Such catastrophic accidents do not happen suddenly; there would be enough time to evacuate personnel from the platform to safety. 2. Victims of such accidents, if any, would die "due to other reasons" long before critical radiation from unshielded canisters could kill them.	θ = E - 6

APPENDIX D DRILLED EMPLACEMENT METHOD

FAULT TREE ANALYSIS - HAZARD TO PERSONNEL

DATA ANALYSIS - EXPLANATORY NOTES

LOGIC REFERENCE	DERIVATION OF FAILURE RATES OR PROBABILITY	ASSESSMENT (θ = f/year) (p = f probability)
B.8	This θ value is based on the safety record of the nuclear industry inspection proceedings and standard with regard to the so called "sniffing method".	$\theta = 5$ E - 5
B.9	From B10, B11, B12 & B13. This o figure, which refers to the number of accidents of critical radiation, that leads to short time death, would give a death risk per year r = 3.33 E - 7. The 'r' value is calculated as follows: Assumed number of deaths per accident: 5. As discussed previously, this is a most pessimistic figure. Thus: $r = \theta \times \frac{5}{500} = 3.33$ E - 7 If death risk is related to wet cell personnel of about 30, then $r = \theta \times \frac{5}{30} = 5.55$ E - 6	$\theta = 3.33$ E - 5

APPENDIX D DRILLED EMPLACEMENT METHOD

FAULT TREE ANALYSIS - HAZARD TO PERSONNEL

DATA ANALYSIS - EXPLANATORY NOTES

LOGIC REFERENCE	DERIVATION OF FAILURE RATES OR PROBABILITY	ASSESSMENT (θ = f/year) (p = f probability)
B.9 Cont.	This figure too, is still well within the NRPB recommendation.	
B.10	From B14, B15 & B16.	$\theta = 1\text{ E} - 5$
B.11	From B17, B18 & B19.	$\theta = 2.2\text{ E} - 5$
B.12	From B20, B21 & B28.	$\theta = 1\text{ E} - 6$
B.13	From B24, B27 & B29.	$\theta = 2.5\text{ E} - 7$
B.14	Personnel could fall into wet cell water through three different routes:- 1. One has to enter wet cell space. Such entry may occur through a command, (i.e. to carry out an inspection or a maintenance function), through a mistake (i.e. someone goes there inadvertently; this, however, will be very difficult as entry is not possible only after a number of safety procedures) and through a suicide attempt. (This, like the one through a mistake, will equally be extremely difficult).	$\theta = 1\text{ E} - 2$

LOGIC REFERENCE	DERIVATION OF FAILURE RATES OR PROBABILITY	ASSESSMENT (θ = f/year) (p = f probability)
B.14 Cont.	2. One could fall in from deck when shielding roof is open leaving a gap through which a person can pass. 3. Finally, one could fall into wet cell water through the shaft which provides a free passage for the flask between the platform deck and wet cell. A fall via routes 2 and 3 may occur for the same three reasons as in route 1, (i.e. accidental fall when men are working in the area; accidental fall when a person inadvertently enters the danger zone; and thirdly when someone attempts a suicide). All above accidents are very unlikely events. Moreover, should such a fall happen, it is very likely that death will occur due to impact, and almost instantaneously, long before a critical radiation could take effect.	
B.15	The wet cell water is contaminated. In this context, contamination means that sufficient radioactive particles are present in the water-mass or the person is too near to the radioactive canisters, i.e. critical radiation reaches the body.	p = 1 E - 1
B.16	The failure of a quick rescue operation leads to the uptake of a critical radiation dose.	p = 1 E - 2

APPENDIX D DRILLED EMPLACEMENT METHOD

FAULT TREE ANALYSIS - HAZARD TO PERSONNEL

DATA ANALYSIS - EXPLANATORY NOTES

LOGIC REFERENCE	DERIVATION OF FAILURE RATES OR PROBABILITY	ASSESSMENT (θ = f/year) (p = f probability)
B.17	It is assumed that when a malfunction occurs and all remote efforts fail to remedy fault, one diver or possibly two will have to enter the wet cell water. This will be carried out, in a protective clothing with the utmost personnel safety enforced. Suicide here is not considered.	p = 2.2 E - 1
B.18	See remarks in B.14.	p = 1E - 1
B.19	Diver working in wet cell water is in instant communication with control room. The radiation reaching him is also monitored. He is ordered out of water when the time is up or when an emergency arose. At present a long enough stay in water is only foreseen if diver is trapped.	p = 1 E - 3
B.20	For conditions of entering wet cell space see explanatory note in B.14.(1)	p = 4E - 1
B.21	Personnel required to enter space are also in visual and sound contact with control room. In any emergency of high level radiation personnel is ordered out of this place. Only if disobeying orders or if exit door systems do malfunction can a long stay	p = 1E - 3

APPENDIX D DRILLED EMPLACEMENT METHOD

FAULT TREE ANALYSIS - HAZARD TO PERSONNEL

DATA ANALYSIS - EXPLANATORY NOTES

LOGIC REFERENCE	DERIVATION OF FAILURE RATES OR PROBABILITY	ASSESSMENT (θ = f/year) (p = f probability)
B.21 Cont.	be foreseen.	
B.22	For figures and explanation see Appendix B, Logic Ref. 66.	p = 5E - 4
B.23	For figures and explanation see Appendix B, Logic Ref. 65.	p = 2E - 2
B.24	Roof shield not fully closed is from Appendix B, Logic Ref. 67.	p = 1E - 3
B.25	Same as B.22.	p = 5E - 4
B.26	Same as B.23.	p = 2E - 3
B.27	Personnel on deck in line of radiation. This can only happen if wet cell shielding roof is open. The figure of one person in ten years, staying long enough is a pessimistic one, but it was used deliberately in order to achieve a safe and acceptable hazard risk.	p = 1E - 1

APPENDIX D DRILLED EMPLACEMENT METHOD

FAULT TREE ANALYSIS - HAZARD TO PERSONNEL

DATA ANALYSIS - EXPLANATORY NOTES

LOGIC REFERENCE	DERIVATION OF FAILURE RATES OR PROBABILITY	ASSESSMENT (Θ = f/year) (p = f probability)
B.28	No shielding (partial or total) by water is calculated from B22 and B23.	p = 2.5 E - 3
B.29	Same as B.28.	p = 2.5 E - 3

APPENDIX D
RADIATION UNITS

Units of Specific Absorbed Dose: Imparting energy to a specific medium, e.g. tissue.

SI Unit is the Gray : 1 Gr = 1 Joule/kg

Rad : 1 Rad = 0.01 Gr = 0.01 Joule/kg

1 Rad = 10^5 ergs/kg = 10^2 ergs/gr

Units of Dose Equivalent (Effective biological dose)

SI Unit is the Sievert 1 Sv = Specified absorbed dose in Grays x QF.

QF is a dimensionaless quantity called the quantity factor which is taken as follows:

QF = 20 for alpha particles in the 1-7 MeV range.

QF = 1 for X, gamma, beta+, beta- emissions.

1 Rontgen equivalent physical (Rep) is the dose which deposits 93 ergs/cm^3 in tissue.

1 Rontgen equivalent man (Rem) is the dose of any radiation which produces the same effect as 1 Rontgen of X-rays or gamma rays. Related to the Rad by a factor termed the relative biological effectiveness.

For X-rays and gamma rays 1 Rem corresponds to the depostion of 0.01 Joule of energy per kilogram of material. Therefore for these radiations:

1 Rem = 1 Rad = 0.01 Gr. = 0.01 Sv; 1 Sv = 100 Rem.

Units of Radioactive Disintegration

SI Unit is the Bequerel : 1 Bq = 1 disintegration/sec.

Curie 1 Ci = 3.7 x 10^{10} Bq.

RADIATION. SOMATIC RISK ONLY

Early effects

(Hereditary effects are not discussed)

The units used to measure doses of radiation are the rem and the millirem (mrem). The mrem is one-thousandth of a rem. Doses of thousands of rem to small regions of the body are used in radiotherapy to destroy cancerous growths. A single dose of about 1000 rem delivered within a few minutes to the whole body, however, is generally fatal.

No early death has been recorded from simple doses of 100 rem or less. Patients exposed to such radiation have generally shown good recovery from loss of hair, from radiation sickness and from skinburn. The risk of early death from a single exposure to radiation of the whole body can be taken as low as zero at 100 rem and less, rising to 100% at 1000 rem and above.

Delayed Effects

The most important delayed somatic effect of radiation is cancer.

DEEP DRILLED EMPLACEMENT - HAZARD TO PERSONNEL

FAULT TREE

10 gates, 21 basic events.

```
B1                           LOSS OF LIFE
       due to: B2                LIVES LOST DUE TO RADIATION
           OR  B3                LIVES LOST DUE TO OPERATIONAL ACTIVITIES
           OR  B4                LIVES LOST DUE TO PLATFORM SINKING
           OR  B5                LIVES LOST DUE TO SHIP SINKING
           OR  B6                ACCIDENTS ON SHIPS (EXCLUDING SINKING)

B2                           LIVES LOST DUE TO RADIATION
       due to: B120              EXPOSURE TO RADIATION
           AND B121              PROPORTION OF PERSONS INVOLVED

B120                         EXPOSURE TO RADIATION
       due to: B7                EXPOSURE TO RADIATION ON DECK
           OR  B8                EXPOSURE TO RADIATION IN INSP. CHAMBER
           OR  B9                EXPOSURE TO RADIATION IN WET CELL

B9                           EXPOSURE TO RADIATION IN WET CELL
       due to: B10               PERSON FALLS IN WET-CELL WATER
           OR  B11               DIVER ENTERS WATER
           OR  B12               PERSON IN WET CELL SPACE
           OR  B13               PERSON OUTSIDE WET CELL

B10                          PERSON FALLS IN WET-CELL WATER
       due to: B14               PERSON FALLS IN
           AND B15               WATER CONTAMINATED
           AND B16               NOT RESCUED FAST ENOUGH

B11                          DIVER ENTERS WATER
       due to: B17               DIVER REQUIRED TO ENTER
           AND B18               WATER CONTAMINATED
           AND B19               DIVER STAYS TOO LONG

B12                          PERSON IN WET CELL SPACE
       due to: B20               PERSON ENTERS WET CELL SPACE
           AND B21               STAYS TOO LONG
           AND B28               NO SHIELD CASE

B28                          NO SHIELD CASE
       due to: B22               CANISTER LIFTED HIGH
           OR  B23               WATER LEVEL FALLS

B13                          PERSON OUTSIDE WET CELL
       due to: B24               ROOF SHIELD OPEN
           AND B27               TOO LONG EXPOSURE
           AND B29               NO SHIELDING FROM WATER

B29                          NO SHIELDING FROM WATER
       due to: B25               CANISTER IS LIFTED HIGH
           OR  B26               WATER LEVEL DROPS
```

DEEP DRILLED EMPLACEMENT - HAZARD TO PERSONNEL

	BASIC EVENTS	FAILURE RATE
B3	LIVES LOST DUE TO OPERATIONAL ACTIVITIES	Ø.62ØE-Ø3
B4	LIVES LOST DUE TO PLATFORM SINKING	Ø.1ØØE-Ø3
B5	LIVES LOST DUE TO SHIP SINKING	Ø.5ØØE-Ø4
B6	ACCIDENTS ON SHIPS (EXCLUDING SINKING)	Ø.46ØE-Ø3
B121	PROPORTION OF PERSONS INVOLVED	Ø.1ØØE-Ø1
B7	EXPOSURE TO RADIATION ON DECK	Ø.1ØØE-Ø5
B8	EXPOSURE TO RADIATION IN INSP. CHAMBER	Ø.5ØØE-Ø4
B14	PERSON FALLS IN	Ø.1ØØE-Ø1
B15	WATER CONTAMINATED	Ø.1ØØE ØØ
B16	NOT RESCUED FAST ENOUGH	Ø.1ØØE-Ø1
B17	DIVER REQUIRED TO ENTER	Ø.22ØE ØØ
B18	WATER CONTAMINATED	Ø.1ØØE ØØ
B19	DIVER STAYS TOO LONG	Ø.1ØØE-Ø2
B2Ø	PERSON ENTERS WET CELL SPACE	Ø.4ØØE ØØ
B21	STAYS TOO LONG	Ø.1ØØE-Ø2
B22	CANISTER LIFTED HIGH	Ø.5ØØE-Ø3
B23	WATER LEVEL FALLS	Ø.2ØØE-Ø2
B24	ROOF SHIELD OPEN	Ø.1ØØE-Ø2
B27	TOO LONG EXPOSURE	Ø.1ØØE ØØ
B25	CANISTER IS LIFTED HIGH	Ø.5ØØE-Ø3
B26	WATER LEVEL DROPS	Ø.2ØØE-Ø2

Probability limit = Ø.ØØØE ØØ   Maximum length of cut set = 1Ø

DEEP DRILLED EMPLACEMENT - HAZARD TO PERSONNEL

Cut set listing for B1 : LOSS OF LIFE
12 minimal cut sets

Probability	Cut set list			
Ø.62ØE-Ø3	B3			
Ø.46ØE-Ø3	B6			
Ø.1ØØE-Ø3	B4			
Ø.5ØØE-Ø4	B5			
Ø.5ØØE-Ø6	B121	B8		
Ø.22ØE-Ø6	B121	B17	B18	B19
Ø.1ØØE-Ø6	B121	B14	B15	B16
Ø.1ØØE-Ø7	B121	B7		
Ø.8ØØE-Ø8	B121	B2Ø	B21	B23
Ø.2ØØE-Ø8	B121	B24	B27	B26
Ø.2ØØE-Ø8	B121	B2Ø	B21	B22
Ø.5ØØE-Ø9	B121	B24	B27	B25

LOSS OF LIFE
OVERALL PROBABILITY OF FAILURE = Ø.123E-Ø2 proportion of staff/year

For a total staff of 5ØØ, annual loss of life = Ø.615 persons/year

Attributable to non-radioactive accidents Ø.615 persons/year
Attributable to radioactive accidents 4.2 E-Ø4 persons/year

DEEP DRILLED EMPLACEMENT - HAZARD TO PERSONNEL

ANALYSIS OF IMPORTANCE

	Basic Event	Failure Rate	Importance
B3	LIVES LOST DUE TO OPERATIONAL ACTIVITIES	Ø.62ØE-Ø3	Ø.5Ø4E ØØ
B6	ACCIDENTS ON SHIPS (EXCLUDING SINKING)	Ø.46ØE-Ø3	Ø.374E ØØ
B4	LIVES LOST DUE TO PLATFORM SINKING	Ø.1ØØE-Ø3	Ø.813E-Ø1
B5	LIVES LOST DUE TO SHIP SINKING	Ø.5ØØE-Ø4	Ø.4Ø6E-Ø1
B121	PROPORTION OF PERSONS INVOLVED	Ø.1ØØE-Ø1	Ø.685E-Ø3
B8	EXPOSURE TO RADIATION IN INSP. CHAMBER	Ø.5ØØE-Ø4	Ø.4Ø6E-Ø3
B17	DIVER REQUIRED TO ENTER	Ø.22ØE ØØ	Ø.179E-Ø3
B18	WATER CONTAMINATED	Ø.1ØØE ØØ	Ø.179E-Ø3
B19	DIVER STAYS TOO LONG	Ø.1ØØE-Ø2	Ø.179E-Ø3
B14	PERSON FALLS IN	Ø.1ØØE-Ø1	Ø.813E-Ø4
B15	WATER CONTAMINATED	Ø.1ØØE ØØ	Ø.813E-Ø4
B16	NOT RESCUED FAST ENOUGH	Ø.1ØØE-Ø1	Ø.813E-Ø4
B7	EXPOSURE TO RADIATION ON DECK	Ø.1ØØE-Ø5	Ø.813E-Ø5
B2Ø	PERSON ENTERS WET CELL SPACE	Ø.4ØØE ØØ	Ø.813E-Ø5
B21	STAYS TOO LONG	Ø.1ØØE-Ø2	Ø.813E-Ø5
B23	WATER LEVEL FALLS	Ø.2ØØE-Ø2	Ø.65ØE-Ø5
B24	ROOF SHIELD OPEN	Ø.1ØØE-Ø2	Ø.2Ø3E-Ø5
B27	TOO LONG EXPOSURE	Ø.1ØØE ØØ	Ø.2Ø3E-Ø5
B26	WATER LEVEL DROPS	Ø.2ØØE-Ø2	Ø.163E-Ø5
B22	CANISTER LIFTED HIGH	Ø.5ØØE-Ø3	Ø.163E-Ø5
B25	CANISTER IS LIFTED HIGH	Ø.5ØØE-Ø3	Ø.4Ø6E-Ø6

CPU time = 2.Ø8 Sec TUE, 18 JUN 1985

TABLE D1

LIVES LOST DUE TO ACCIDENTS WHILE PERSONNEL ON SHIP DURING VOYAGE TO & FROM PLATFORM

The following data taken from Table 20 of "CASUALTIES TO VESSELS & ACCIDENTS TO MEN 1981". This publication is produced yearly by the Department of Trade and Department of Transport and is obtainable from HMSO.

A. Death from Casualties to Vessel	YEAR						
	1975	1976	1977	1978	1979	1980	1975 to 1980
Founderings	24				12	1	37
Collisions	1			5			6
Explosion & Fires	7	5	3	3	4	2	24
Other Casualties	1	1			3		5
TOTAL	33	6	3	8	19	3	72
TOTAL WITHOUT FOUNDERINGS	9				7	2	35
B. Death from accidents other than casualties to vessel							
Accidents on Board	38	32	37	63	28	22	220
TOTAL (A+B)	71	38	40	71	47	25	292
TOTAL (A+B) WITHOUT FOUNDERINGS	47				35	24	255
Estimated number of Seamen at risk (thousands)	108	103	95	88	82	78	554

Death per 100 Seamen p.a $\quad d = 100 : \dfrac{554000}{292} = 5.3\ E-2$

Death per 100 Seamen p.a without founderings $\quad d = 100 : \dfrac{554000}{255} = 4.6\ E-2$

TABLE D2

Average annual risk of death in the UK from accidents in various industries and from cancers potentially induced among radiation workers

Industries	Risk of Death per year	(y^{-1})
Deep Sea fishing	1 in 400	(2.5×10^{-3})
Coal mining	1 in 4000	(2.5×10^{-4})
Construction	1 in 5000	(2.0×10^{-4})
Metal Manufacture	1 in 7000	(1.4×10^{-4})
Timber, Furniture etc	1 in 17,000	(5.9×10^{-5})
All employment	1 in 20,000	(5.0×10^{-5})
Radiation Workers (4mSv per year average)	1 in 20,000	(5.0×10^{-5})
Food, Drink and Tobacco	1 in 30,000	(3.3×10^{-5})
Textiles	1 in 40,000	(2.5×10^{-5})
Clothing & Footwear	1 in 300,000	(3.3×10^{-6})

Average annual risk of death in the UK from some common causes and from cancers potentially induced among highly-exposed individuals

Cause	Risk of Death per year	(y^{-1})
Smoking 10 cigarettes a day	1 in 200	(5×10^{-3})
Natural Causes, 40 years old	1 in 500	(2×10^{-3})
Accidents on the road	1 in 5000	(2×10^{-4})
Accidents in the home	1 in 10,000	(1×10^{-4})
Accidents at work	1 in 20,000	(5×10^{-5})
Radiation exposure (1mSv per year)	1 in 80,000	(1.25×10^{-5})

TABLE D-2 ANNUAL RISKS OF DEATH. GENERAL STATISTICAL DATA

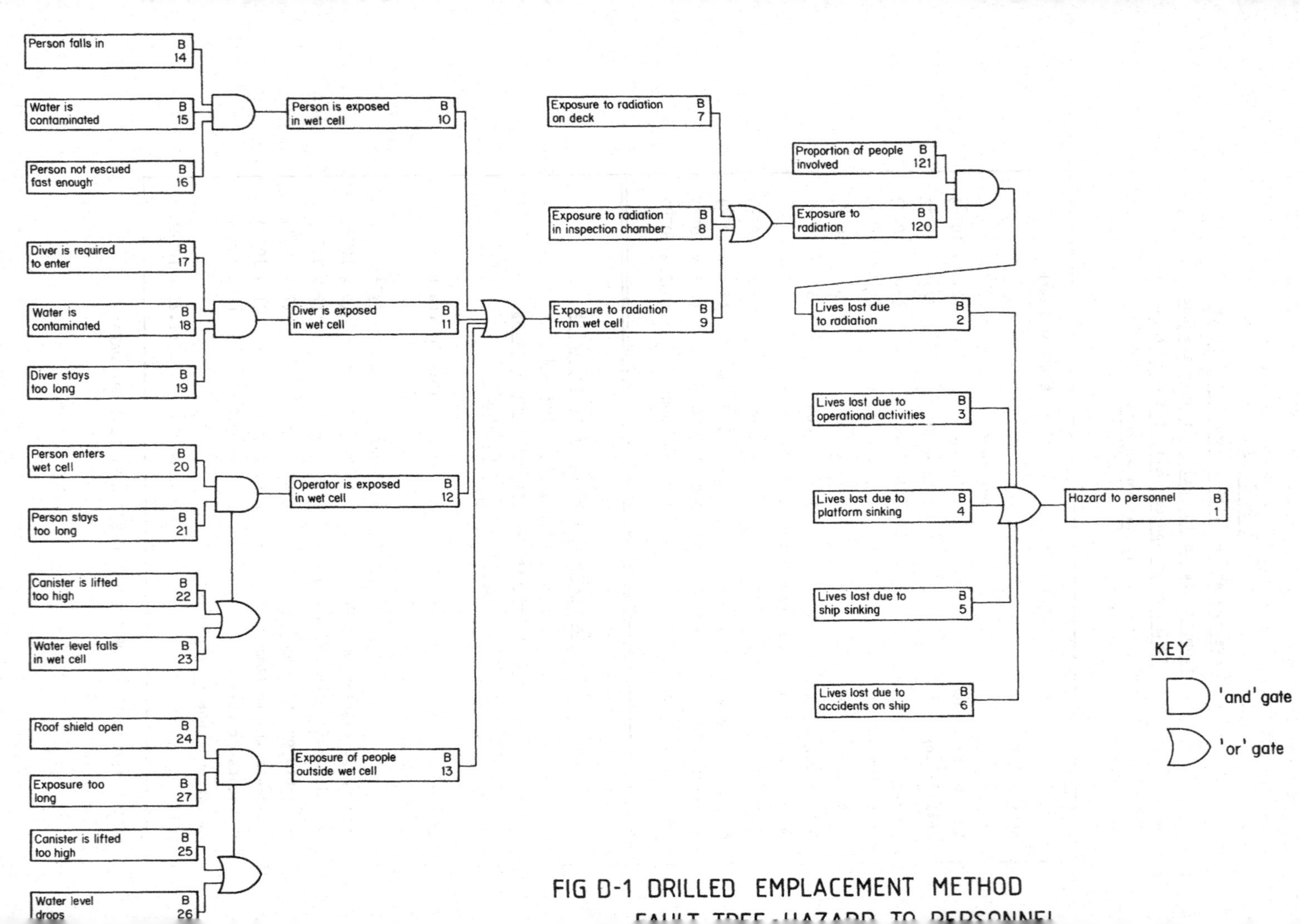

FIG D-1 DRILLED EMPLACEMENT METHOD

APPENDIX E

FAULT TREE ANALYSIS

'RELEASE OF RADIOACTIVITY TO THE ENVIRONMENT'

PENETRATOR METHOD

CONTENTS

APPENDIX E FAULT TREE ANALYSIS - PENETRATOR METHOD

DATA ANALYSIS - EXPLANATORY NOTES

LOGIC REFERENCE	DERIVATION OF FAILURE RATES OR PROBABILITY	ASSESSMENT (θ = f/year) (p = f probability)
P1	From Fault Tree Logic	
P2	From Fault Tree Logic	
P3, P4	The penetrator flask will be designed and tested to ensure that it will be able to withstand the maximum thermal and dynamic conditions to which it can possibly be subjected. Tests on other transport flasks have indicated that no breach will occur under these conditions.	$\theta = 5 \times 10^{-6}$
P6	Statistics provided by Lloyds Register indicate that 0.33% of the World's cargo fleets are lost annually as a result of being stranded or being involved in collisions, most of which would be in congested coastal water. This figure covers a wide range of ships and navigating equipment. A figure of 0.0015 per ship is taken, assuming 3 ships, shared between P6 and P11 in ratio 1:2.	$\theta = 1.5 \times 10^{-3}$
P7	For the preferred option, the transport flask is loaded onto the disposal ship using a deck crane. The incidence that a crane fault or human error would allow a flask to fall is assessed as 5×10^{-4} per year.	$\theta = 5 \times 10^{-4}$
P8	The possibility that a flask could not be recovered from shallow water in the harbour is considered to be low in view of the successful recoveries of materials from much more difficult situations.	$\theta = 1 \times 10^{-3}$

APPENDIX E FAULT TREE ANALYSIS - PENETRATOR METHOD

DATA ANALYSIS - EXPLANATORY NOTES

LOGIC REFERENCE	DERIVATION OF FAILURE RATES OR PROBABILITY	ASSESSMENT (θ = f/year) (p = f probability)
P9	Statistics provided by Lloyds Register indicates that 0.38% of the World's cargo fleets are lost in the open sea each year. Most of these lossess are due to foundering after hull failure, cargo shift or leakage through latches or machinery pipelines. A large proportion of the crafts are fishing vessels and small crafts, a major contributory factor being neglected maintenance. It is noted that very few vessels are actually sunk by fire damage. The vessels used for transporting flasks will be designed, built and maintained to high standards. A figure of 0.0005 per ship is taken, assuming 3 ships.	$\theta = 1.5 \times 10^{-3}$
P10	It is anticipated that the penetrator will incorporate a feature to assist location under water. It should therefore be possible to ascertain the position even if one has broken away from a sunken ship. The probability of failure to recover all penetrators sunk with a transport ship is a function of failure to locate (feature failed or inefficient for various reasons) and failure to recover (impossible situation) before penetrators were breached by corrosion. The water depth could be between 200 and 5500m. The probabilities are (somewhat subjectively) assessed as: Failure to locate every penetrator 0.5 Failure to recover every penetrator after location 0.5 Failure to locate OR recover = 1 - (0.5 x 0.5)	p = 0.75 to p = 0.999

APPENDIX E FAULT TREE ANALYSIS - PENETRATOR METHOD

DATA ANALYSIS - EXPLANATORY NOTES

LOGIC REFERENCE	DERIVATION OF FAILURE RATES OR PROBABILITY	ASSESSMENT (θ = f/year) (p = f probability)
P10 Cont.	p = 0.75 is somewhat optimistic for deep water conditions and a value of 0.999 has been taken as a pessimistic value to see the effect on the overall probability value.	
P11	See Reference P6 above	$\theta = 3 \times 10^{-3}$
P12	Recovery of penetrators from shallow water is expected to be much easier than for deeper water.	p = 0.02
P13	From Fault Tree Logic.	
P14	Recovery of unsuccessfully embedded penetrator is considered to be remote.	p = 0.999
P15	From Fault Tree Logic	
P16	Mechanical malfunction leading to accidental release is low for most systems, but could be occasioned by for example testing the system for deployment. Assessment given is for Option E, but could be much higher for Option B with multiple release parts.	$\theta = 1 \times 10^{-4}$
P17	Human error leading to accidental release is considered on similar ground to Reference P16 above.	$\theta = 1 \times 10^{-4}$

APPENDIX E FAULT TREE ANALYSIS - PENETRATOR METHOD

DATA ANALYSIS - EXPLANATORY NOTES

LOGIC REFERENCE	DERIVATION OF FAILURE RATES OR PROBABILITY	ASSESSMENT (θ = f/year) (p = f probability)
P19	From Fault Tree Logic	
P20	The nominal launch spacing has been estimated using combined available information on emplaced waste spacing, positional accuracy of the disposal ship, penetrator trajectories in still water and ocean current allowances. The estimated launch spacing of 125m is over eight times greater than the design emplacement spacing of 15. The possibilities of penetrators ending up very close to each other is considered to be low.	$\theta = 1 \times 10^{-4}$
P21	Theoretical predictions carried out by OAP indicate that the designed speed is achievable. A sensitivity study carried out for various misalignments on the reference penetrator indicated negligible effects on the terminal velocity of the penetrator. The assumed failure rate taken to be 2 in a million penetrators.	$\theta = 1 \times 10^{-3}$
P22	The possibility of a penetrator hitting a glacial erratic or any other solid obstruction was estimated to be very low. The OAP study indicated that the probability of an impact with a 0.2m diameter boulder buried 0.7m below the surface was estimated to be 3×10^{-5}, and say this is unlikely to affect penetration depth. Chance of hitting a sufficiently large boulder is therefore estimated to be 2×10^{-7} per penetrator.	$\theta = 1 \times 10^{-4}$

APPENDIX E FAULT TREE ANALYSIS - PENETRATOR METHOD

DATA ANALYSIS - EXPLANATORY NOTES

LOGIC REFERENCE	DERIVATION OF FAILURE RATES OR PROBABILITY	ASSESSMENT (θ = f/year) (p = f probability)
P23	Investigations by the Nuclear Energy Agency Seabed Working Group have so far not found any information that the seabed may have a greater resistance than has been designed for. It is anticipated that a dummy penetrator will be dropped first, and subsequent penetrator depths will be monitored. Any indication that the sediments become stiffer will be noticed and further releases halted.	$\theta = 1 \times 10^{-5}$
P24	From Fault Tree Logic	
P25	Penetrator imperfections could be caused by manufacturing faults, damage to fins during handling, non-axisymmetry etc. Assuming a damage rate to fins of 2×10^{-5} per penetrator, the annual failure rate is 1×10^{-2} per year. The damage has to be sufficient to cause the penetrator to enter the sediment at a sufficient angle to cause it not to reach its design depth.	$\theta = 1 \times 10^{-2}$
P26	Ocean current data has been obtained from the Institute of Oceanographic Sciences. This data is included in the OAP analysis by increasing the flow in the horizontal direction by a value appropriate to the depth of the penetrator. Ocean currents effect on the displacement is estimated to be of the order of 5-10m, with heavier and hence faster penetrators being less affected.	$\theta = 1 \times 10^{-4}$

APPENDIX E FAULT TREE ANALYSIS - PENETRATOR METHOD

DATA ANALYSIS - EXPLANATORY NOTES

LOGIC REFERENCE	DERIVATION OF FAILURE RATES OR PROBABILITY	ASSESSMENT (θ = f/year) (p = f probability)
P27	The OAP computer model predicted that the reference penetrator would not deviate significantly from its attitude at entry and the required embedment depths would always be achieved. Owing to lack of experimental observations and the uncertainty involved in applying some of the geotechnical theory, it would not be rewarding to examine this failure in any detail at this early stage of development.	$\theta = 1 \times 10^{-4}$

PENETRATOR SYSTEM
CASE 1 - NO RECOVERY FROM DEEP WATER

FAULT TREE

11 gates, 19 basic events.

Gate	Relation	Event	Description
P1			RELEASE OF RADIOACTIVITY TO ENVIRONMENT
	due to:	P2	PENETRATOR LOST IN SEA
	OR	P3	PENETRATOR BREACHED BY IMPACT
	OR	P4	PENETRATOR BREACHED BY FIRE
P2			PENETRATOR LOST IN SEA
	due to:	P121	PENETRATOR LOST IN HARBOUR
	OR	P122	PENETRATOR LOST IN OPEN SEA
	OR	P123	PENETRATOR LOST NEAR SHORE
	OR	P124	PENETRATOR MISPLACED AND LOST
	OR	P15	ACCIDENTAL RELEASE OUTSIDE AREA
P121			PENETRATOR LOST IN HARBOUR
	due to:	P125	PENETRATOR SINKS IN HARBOUR
	AND	P8	RECOVERY IMPOSSIBLE FROM HARBOUR
P125			PENETRATOR SINKS IN HARBOUR
	due to:	P6	SHIP SINKS IN HARBOUR
	OR	P7	PENETRATOR DROPPED IN HARBOUR
P122			PENETRATOR LOST IN OPEN SEA
	due to:	P9	SHIP SINKS IN OPEN SEA
	AND	P1Ø	RECOVERY IMPOSSIBLE FROM OPEN SEA
P123			PENETRATOR LOST NEAR SHORE
	due to:	P11	SHIP SINKS NEAR SHORE
	AND	P12	RECOVERY IMPOSSIBLE FROM SHALLOW WATER
P124			PENETRATOR MISPLACED AND LOST
	due to:	P13	PENETRATOR UNSUCCESSFULLY EMBEDDED
	AND	P14	MISPLACED PENETRATOR NOT RECOVERED
P13			PENETRATOR UNSUCCESSFULLY EMBEDDED
	due to:	P19	INSUFFICIENT DEPTH OF EMBEDMENT
	OR	P2Ø	PENETRATOR TOO CLOSE TO ANOTHER
P15			ACCIDENTAL RELEASE OUTSIDE AREA
	due to:	P16	RELEASE DUE TO MECHANICAL MALFUNCTION
	OR	P17	RELEASE DUE TO HUMAN ERROR
P19			INSUFFICIENT DEPTH OF EMBEDMENT
	due to:	P21	PENETRATOR DOES NOT REACH DESIGN SPEED
	OR	P22	PENETRATOR HITS LARGE BOULDER
	OR	P23	SEABED HAS GREAT RESISTANCE
	OR	P24	PENETRATOR DEVIATES THROUGH SEDIMENTS
P24			PENETRATOR DEVIATES THROUGH SEDIMENTS
	due to:	P25	PENETRATOR IMPERFECTIONS (FINS BENT)
	OR	P26	CURRENT VARIABILITY DIVERTS PENETRATOR
	OR	P27	PATH DEVIATIONS DUE TO UNEVEN SEDIMENTS

PENETRATOR SYSTEM
CASE 1 - NO RECOVERY FROM DEEP WATER

	BASIC EVENTS	FAILURE RATE
P3	PENETRATOR BREACHED BY IMPACT	Ø.5ØØE-Ø5
P4	PENETRATOR BREACHED BY FIRE	Ø.5ØØE-Ø5
P6	SHIP SINKS IN HARBOUR	Ø.15ØE-Ø2
P7	PENETRATOR DROPPED IN HARBOUR	Ø.5ØØE-Ø3
P8	RECOVERY IMPOSSIBLE FROM HARBOUR	Ø.1ØØE-Ø2
P9	SHIP SINKS IN OPEN SEA	Ø.15ØE-Ø2
P1Ø	RECOVERY IMPOSSIBLE FROM OPEN SEA	Ø.999E ØØ
P11	SHIP SINKS NEAR SHORE	Ø.3ØØE-Ø2
P12	RECOVERY IMPOSSIBLE FROM SHALLOW WATER	Ø.2ØØE-Ø1
P14	MISPLACED PENETRATOR NOT RECOVERED	Ø.999E ØØ
P16	RELEASE DUE TO MECHANICAL MALFUNCTION	Ø.1ØØE-Ø3
P17	RELEASE DUE TO HUMAN ERROR	Ø.1ØØE-Ø3
P2Ø	PENETRATOR TOO CLOSE TO ANOTHER	Ø.1ØØE-Ø3
P21	PENETRATOR DOES NOT REACH DESIGN SPEED	Ø.1ØØE-Ø2
P22	PENETRATOR HITS LARGE BOULDER	Ø.1ØØE-Ø3
P23	SEABED HAS GREAT RESISTANCE	Ø.1ØØE-Ø4
P25	PENETRATOR IMPERFECTIONS (FINS BENT)	Ø.1ØØE-Ø1
P26	CURRENT VARIABILITY DIVERTS PENETRATOR	Ø.1ØØE-Ø3
P27	PATH DEVIATIONS DUE TO UNEVEN SEDIMENTS	Ø.1ØØE-Ø3

Probability limit = Ø.ØØØE ØØ Maximum length of cut set = 1Ø

PENETRATOR SYSTEM
CASE 1 - NO RECOVERY FROM DEEP WATER

Cut set listing for P1 : RELEASE OF RADIOACTIVITY TO ENVIRONMENT
15 minimal cut sets

Probability	Cut set list	
0.999E-02	P14	P25
0.150E-02	P9	P10
0.999E-03	P14	P21
0.100E-03	P16	
0.100E-03	P17	
0.999E-04	P14	P27
0.999E-04	P14	P20
0.999E-04	P14	P22
0.999E-04	P14	P26
0.600E-04	P11	P12
0.999E-05	P14	P23
0.500E-05	P3	
0.500E-05	P4	
0.150E-05	P6	P8
0.500E-06	P7	P8

OVERALL PROBABILITY OF FAILURE = 0.131E-01

PENETRATOR SYSTEM
CASE 1 - NO RECOVERY FROM DEEP WATER

ANALYSIS OF IMPORTANCE

	Basic Event	Failure Rate	Importance
P14	MISPLACED PENETRATOR NOT RECOVERED	Ø.999E ØØ	Ø.867E ØØ
P25	PENETRATOR IMPERFECTIONS (FINS BENT)	Ø.1ØØE-Ø1	Ø.761E ØØ
P9	SHIP SINKS IN OPEN SEA	Ø.15ØE-Ø2	Ø.114E ØØ
P1Ø	RECOVERY IMPOSSIBLE FROM OPEN SEA	Ø.999E ØØ	Ø.114E ØØ
P21	PENETRATOR DOES NOT REACH DESIGN SPEED	Ø.1ØØE-Ø2	Ø.761E-Ø1
P16	RELEASE DUE TO MECHANICAL MALFUNCTION	Ø.1ØØE-Ø3	Ø.761E-Ø2
P17	RELEASE DUE TO HUMAN ERROR	Ø.1ØØE-Ø3	Ø.761E-Ø2
P22	PENETRATOR HITS LARGE BOULDER	Ø.1ØØE-Ø3	Ø.761E-Ø2
P2Ø	PENETRATOR TOO CLOSE TO ANOTHER	Ø.1ØØE-Ø3	Ø.761E-Ø2
P26	CURRENT VARIABILITY DIVERTS PENETRATOR	Ø.1ØØE-Ø3	Ø.761E-Ø2
P27	PATH DEVIATIONS DUE TO UNEVEN SEDIMENTS	Ø.1ØØE-Ø3	Ø.761E-Ø2
P12	RECOVERY IMPOSSIBLE FROM SHALLOW WATER	Ø.2ØØE-Ø1	Ø.457E-Ø2
P11	SHIP SINKS NEAR SHORE	Ø.3ØØE-Ø2	Ø.457E-Ø2
P23	SEABED HAS GREAT RESISTANCE	Ø.1ØØE-Ø4	Ø.761E-Ø3
P3	PENETRATOR BREACHED BY IMPACT	Ø.5ØØE-Ø5	Ø.381E-Ø3
P4	PENETRATOR BREACHED BY FIRE	Ø.5ØØE-Ø5	Ø.381E-Ø3
P8	RECOVERY IMPOSSIBLE FROM HARBOUR	Ø.1ØØE-Ø2	Ø.152E-Ø3
P6	SHIP SINKS IN HARBOUR	Ø.15ØE-Ø2	Ø.114E-Ø3
P7	PENETRATOR DROPPED IN HARBOUR	Ø.5ØØE-Ø3	Ø.381E-Ø4

CPU time = 2.Ø3 Sec TUE, 17 SEP 1985

PENETRATOR SYSTEM
CASE 2 - SOME RECOVERY FROM DEEP WATER

FAULT TREE

11 gates, 19 basic events.

```
P1                    RELEASE OF RADIOACTIVITY TO ENVIRONMENT
      due to: P2          PENETRATOR LOST IN SEA
           OR P3          PENETRATOR BREACHED BY IMPACT
           OR P4          PENETRATOR BREACHED BY FIRE

P2                    PENETRATOR LOST IN SEA
      due to: P121        PENETRATOR LOST IN HARBOUR
           OR P122        PENETRATOR LOST IN OPEN SEA
           OR P123        PENETRATOR LOST NEAR SHORE
           OR P124        PENETRATOR MISPLACED AND LOST
           OR P15         ACCIDENTAL RELEASE OUTSIDE AREA

P121                  PENETRATOR LOST IN HARBOUR
      due to: P125        PENETRATOR SINKS IN HARBOUR
          AND P8          RECOVERY IMPOSSIBLE FROM HARBOUR

P125                  PENETRATOR SINKS IN HARBOUR
      due to: P6          SHIP SINKS IN HARBOUR
           OR P7          PENETRATOR DROPPED IN HARBOUR

P122                  PENETRATOR LOST IN OPEN SEA
      due to: P9          SHIP SINKS IN OPEN SEA
          AND P10         RECOVERY IMPOSSIBLE FROM OPEN SEA

P123                  PENETRATOR LOST NEAR SHORE
      due to: P11         SHIP SINKS NEAR SHORE
          AND P12         RECOVERY IMPOSSIBLE FROM SHALLOW WATER

P124                  PENETRATOR MISPLACED AND LOST
      due to: P13         PENETRATOR UNSUCCESSFULLY EMBEDDED
          AND P14         MISPLACED PENETRATOR NOT RECOVERED

P13                   PENETRATOR UNSUCCESSFULLY EMBEDDED
      due to: P19         INSUFFICIENT DEPTH OF EMBEDMENT
           OR P20         PENETRATOR TOO CLOSE TO ANOTHER

P15                   ACCIDENTAL RELEASE OUTSIDE AREA
      due to: P16         RELEASE DUE TO MECHANICAL MALFUNCTION
           OR P17         RELEASE DUE TO HUMAN ERROR

P19                   INSUFFICIENT DEPTH OF EMBEDMENT
      due to: P21         PENETRATOR DOES NOT REACH DESIGN SPEED
           OR P22         PENETRATOR HITS LARGE BOULDER
           OR P23         SEABED HAS GREAT RESISTANCE
           OR P24         PENETRATOR DEVIATES THROUGH SEDIMENTS

P24                   PENETRATOR DEVIATES THROUGH SEDIMENTS
      due to: P25         PENETRATOR IMPERFECTIONS (FINS BENT)
           OR P26         CURRENT VARIABILITY DIVERTS PENETRATOR
           OR P27         PATH DEVIATIONS DUE TO UNEVEN SEDIMENTS
```

PENETRATOR SYSTEM
CASE 2 - SOME RECOVERY FROM DEEP WATER

	BASIC EVENTS	FAILURE RATE
P3	PENETRATOR BREACHED BY IMPACT	Ø.5ØØE-Ø5
P4	PENETRATOR BREACHED BY FIRE	Ø.5ØØE-Ø5
P6	SHIP SINKS IN HARBOUR	Ø.15ØE-Ø2
P7	PENETRATOR DROPPED IN HARBOUR	Ø.5ØØE-Ø3
P8	RECOVERY IMPOSSIBLE FROM HARBOUR	Ø.1ØØE-Ø2
P9	SHIP SINKS IN OPEN SEA	Ø.15ØE-Ø2
P1Ø	RECOVERY IMPOSSIBLE FROM OPEN SEA	Ø.75ØE ØØ
P11	SHIP SINKS NEAR SHORE	Ø.3ØØE-Ø2
P12	RECOVERY IMPOSSIBLE FROM SHALLOW WATER	Ø.2ØØE-Ø1
P14	MISPLACED PENETRATOR NOT RECOVERED	Ø.95ØE ØØ
P16	RELEASE DUE TO MECHANICAL MALFUNCTION	Ø.1ØØE-Ø3
P17	RELEASE DUE TO HUMAN ERROR	Ø.1ØØE-Ø3
P2Ø	PENETRATOR TOO CLOSE TO ANOTHER	Ø.1ØØE-Ø3
P21	PENETRATOR DOES NOT REACH DESIGN SPEED	Ø.1ØØE-Ø2
P22	PENETRATOR HITS LARGE BOULDER	Ø.1ØØE-Ø3
P23	SEABED HAS GREAT RESISTANCE	Ø.1ØØE-Ø4
P25	PENETRATOR IMPERFECTIONS (FINS BENT)	Ø.1ØØE-Ø1
P26	CURRENT VARIABILITY DIVERTS PENETRATOR	Ø.1ØØE-Ø3
P27	PATH DEVIATIONS DUE TO UNEVEN SEDIMENTS	Ø.1ØØE-Ø3

Probability limit = Ø.ØØØE ØØ   Maximum length of cut set = 1Ø

PENETRATOR SYSTEM
CASE 2 - SOME RECOVERY FROM DEEP WATER

Cut set listing for P1 : RELEASE OF RADIOACTIVITY TO ENVIRONMENT
15 minimal cut sets

Probability	Cut set list	
Ø.95ØE-Ø2	P14	P25
Ø.112E-Ø2	P9	P1Ø
Ø.95ØE-Ø3	P14	P21
Ø.1ØØE-Ø3	P16	
Ø.1ØØE-Ø3	P17	
Ø.95ØE-Ø4	P14	P27
Ø.95ØE-Ø4	P14	P2Ø
Ø.95ØE-Ø4	P14	P22
Ø.95ØE-Ø4	P14	P26
Ø.6ØØE-Ø4	P11	P12
Ø.95ØE-Ø5	P14	P23
Ø.5ØØE-Ø5	P3	
Ø.5ØØE-Ø5	P4	
Ø.15ØE-Ø5	P6	P8
Ø.5ØØE-Ø6	P7	P8

OVERALL PROBABILITY OF FAILURE = Ø.122E-Ø1

PENETRATOR SYSTEM
CASE 2 - SOME RECOVERY FROM DEEP WATER

ANALYSIS OF IMPORTANCE

	Basic Event	Failure Rate	Importance
P14	MISPLACED PENETRATOR NOT RECOVERED	0.950E 00	0.887E 00
P25	PENETRATOR IMPERFECTIONS (FINS BENT)	0.100E-01	0.778E 00
P9	SHIP SINKS IN OPEN SEA	0.150E-02	0.922E-01
P10	RECOVERY IMPOSSIBLE FROM OPEN SEA	0.750E 00	0.922E-01
P21	PENETRATOR DOES NOT REACH DESIGN SPEED	0.100E-02	0.778E-01
P16	RELEASE DUE TO MECHANICAL MALFUNCTION	0.100E-03	0.819E-02
P17	RELEASE DUE TO HUMAN ERROR	0.100E-03	0.819E-02
P22	PENETRATOR HITS LARGE BOULDER	0.100E-03	0.778E-02
P20	PENETRATOR TOO CLOSE TO ANOTHER	0.100E-03	0.778E-02
P26	CURRENT VARIABILITY DIVERTS PENETRATOR	0.100E-03	0.778E-02
P27	PATH DEVIATIONS DUE TO UNEVEN SEDIMENTS	0.100E-03	0.778E-02
P12	RECOVERY IMPOSSIBLE FROM SHALLOW WATER	0.200E-01	0.491E-02
P11	SHIP SINKS NEAR SHORE	0.300E-02	0.491E-02
P23	SEABED HAS GREAT RESISTANCE	0.100E-04	0.778E-03
P3	PENETRATOR BREACHED BY IMPACT	0.500E-05	0.410E-03
P4	PENETRATOR BREACHED BY FIRE	0.500E-05	0.410E-03
P8	RECOVERY IMPOSSIBLE FROM HARBOUR	0.100E-02	0.164E-03
P6	SHIP SINKS IN HARBOUR	0.150E-02	0.123E-03
P7	PENETRATOR DROPPED IN HARBOUR	0.500E-03	0.410E-04

CPU time = 2.06 Sec TUE, 17 SEP 1985

PENETRATOR SYSTEM
CASE 3 - NO RECOVERY FROM DEEP WATER, MAGNITUDE INCLUDED

FAULT TREE

11 gates, 19 basic events.

```
P1                      RELEASE OF RADIOACTIVITY TO ENVIRONMENT
      due to: P2            PENETRATOR LOST IN SEA
           OR P3            PENETRATOR BREACHED BY IMPACT
           OR P4            PENETRATOR BREACHED BY FIRE

P2                      PENETRATOR LOST IN SEA
      due to: P121          PENETRATOR LOST IN HARBOUR
           OR P122          PENETRATOR LOST IN OPEN SEA
           OR P123          PENETRATOR LOST NEAR SHORE
           OR P124          PENETRATOR MISPLACED AND LOST
           OR P15           ACCIDENTAL RELEASE OUTSIDE AREA

P121                    PENETRATOR LOST IN HARBOUR
      due to: P125          PENETRATOR SINKS IN HARBOUR
          AND P8            RECOVERY IMPOSSIBLE FROM HARBOUR

P125                    PENETRATOR SINKS IN HARBOUR
      due to: P6            SHIP SINKS IN HARBOUR
           OR P7            PENETRATOR DROPPED IN HARBOUR

P122                    PENETRATOR LOST IN OPEN SEA
      due to: P9            SHIP SINKS IN OPEN SEA
          AND P10           RECOVERY IMPOSSIBLE FROM OPEN SEA

P123                    PENETRATOR LOST NEAR SHORE
      due to: P11           SHIP SINKS NEAR SHORE
          AND P12           RECOVERY IMPOSSIBLE FROM SHALLOW WATER

P124                    PENETRATOR MISPLACED AND LOST
      due to: P13           PENETRATOR UNSUCCESSFULLY EMBEDDED
          AND P14           MISPLACED PENETRATOR NOT RECOVERED

P13                     PENETRATOR UNSUCCESSFULLY EMBEDDED
      due to: P19           INSUFFICIENT DEPTH OF EMBEDMENT
           OR P20           PENETRATOR TOO CLOSE TO ANOTHER

P15                     ACCIDENTAL RELEASE OUTSIDE AREA
      due to: P16           RELEASE DUE TO MECHANICAL MALFUNCTION
           OR P17           RELEASE DUE TO HUMAN ERROR

P19                     INSUFFICIENT DEPTH OF EMBEDMENT
      due to: P21           PENETRATOR DOES NOT REACH DESIGN SPEED
           OR P22           PENETRATOR HITS LARGE BOULDER
           OR P23           SEABED HAS GREAT RESISTANCE
           OR P24           PENETRATOR DEVIATES THROUGH SEDIMENTS

P24                     PENETRATOR DEVIATES THROUGH SEDIMENTS
      due to: P25           PENETRATOR IMPERFECTIONS (FINS BENT)
           OR P26           CURRENT VARIABILITY DIVERTS PENETRATOR
           OR P27           PATH DEVIATIONS DUE TO UNEVEN SEDIMENTS
```

PENETRATOR SYSTEM

CASE 3 - NO RECOVERY FROM DEEP WATER, MAGNITUDE INCLUDED

	BASIC EVENTS	FAILURE RATE
P3	PENETRATOR BREACHED BY IMPACT	0.100E-07
P4	PENETRATOR BREACHED BY FIRE	0.500E-06
P6	SHIP SINKS IN HARBOUR	0.500E-04
P7	PENETRATOR DROPPED IN HARBOUR	0.100E-05
P8	RECOVERY IMPOSSIBLE FROM HARBOUR	0.100E-02
P9	SHIP SINKS IN OPEN SEA	0.500E-04
P10	RECOVERY IMPOSSIBLE FROM OPEN SEA	0.999E 00
P11	SHIP SINKS NEAR SHORE	0.100E-03
P12	RECOVERY IMPOSSIBLE FROM SHALLOW WATER	0.200E-01
P14	MISPLACED PENETRATOR NOT RECOVERED	0.999E 00
P16	RELEASE DUE TO MECHANICAL MALFUNCTION	0.200E-06
P17	RELEASE DUE TO HUMAN ERROR	0.200E-06
P20	PENETRATOR TOO CLOSE TO ANOTHER	0.200E-06
P21	PENETRATOR DOES NOT REACH DESIGN SPEED	0.200E-05
P22	PENETRATOR HITS LARGE BOULDER	0.200E-06
P23	SEABED HAS GREAT RESISTANCE	0.200E-07
P25	PENETRATOR IMPERFECTIONS (FINS BENT)	0.200E-04
P26	CURRENT VARIABILITY DIVERTS PENETRATOR	0.200E-06
P27	PATH DEVIATIONS DUE TO UNEVEN SEDIMENTS	0.200E-06

Probability limit = 0.000E 00 Maximum length of cut set = 10

PENETRATOR SYSTEM
CASE 3 - NO RECOVERY FROM DEEP WATER, MAGNITUDE INCLUDED

Cut set listing for P1 : RELEASE OF RADIOACTIVITY TO ENVIRONMENT
15 minimal cut sets

Probability	Cut set list	
Ø.499E-Ø4	P9	P1Ø
Ø.2ØØE-Ø4	P14	P25
Ø.2ØØE-Ø5	P11	P12
Ø.2ØØE-Ø5	P14	P21
Ø.5ØØE-Ø6	P4	
Ø.2ØØE-Ø6	P17	
Ø.2ØØE-Ø6	P16	
Ø.2ØØE-Ø6	P14	P2Ø
Ø.2ØØE-Ø6	P14	P22
Ø.2ØØE-Ø6	P14	P26
Ø.2ØØE-Ø6	P14	P27
Ø.5ØØE-Ø7	P6	P8
Ø.2ØØE-Ø7	P14	P23
Ø.1ØØE-Ø7	P3	
Ø.1ØØE-Ø8	P7	P8

OVERALL PROBABILITY OF FAILURE = Ø.757E-Ø4
= Ø.2 canisters/year

PENETRATOR SYSTEM

CASE 3 - NO RECOVERY FROM DEEP WATER, MAGNITUDE INCLUDED

ANALYSIS OF IMPORTANCE

	Basic Event	Failure Rate	Importance
P9	SHIP SINKS IN OPEN SEA	0.500E-04	0.660E 00
P10	RECOVERY IMPOSSIBLE FROM OPEN SEA	0.999E 00	0.660E 00
P14	MISPLACED PENETRATOR NOT RECOVERED	0.999E 00	0.301E 00
P25	PENETRATOR IMPERFECTIONS (FINS BENT)	0.200E-04	0.264E 00
P11	SHIP SINKS NEAR SHORE	0.100E-03	0.264E-01
P12	RECOVERY IMPOSSIBLE FROM SHALLOW WATER	0.200E-01	0.264E-01
P21	PENETRATOR DOES NOT REACH DESIGN SPEED	0.200E-05	0.264E-01
P4	PENETRATOR BREACHED BY FIRE	0.500E-06	0.660E-02
P16	RELEASE DUE TO MECHANICAL MALFUNCTION	0.200E-06	0.264E-02
P17	RELEASE DUE TO HUMAN ERROR	0.200E-06	0.264E-02
P22	PENETRATOR HITS LARGE BOULDER	0.200E-06	0.264E-02
P20	PENETRATOR TOO CLOSE TO ANOTHER	0.200E-06	0.264E-02
P26	CURRENT VARIABILITY DIVERTS PENETRATOR	0.200E-06	0.264E-02
P27	PATH DEVIATIONS DUE TO UNEVEN SEDIMENTS	0.200E-06	0.264E-02
P8	RECOVERY IMPOSSIBLE FROM HARBOUR	0.100E-02	0.674E-03
P6	SHIP SINKS IN HARBOUR	0.500E-04	0.660E-03
P23	SEABED HAS GREAT RESISTANCE	0.200E-07	0.264E-03
P3	PENETRATOR BREACHED BY IMPACT	0.100E-07	0.132E-03
P7	PENETRATOR DROPPED IN HARBOUR	0.100E-05	0.132E-04

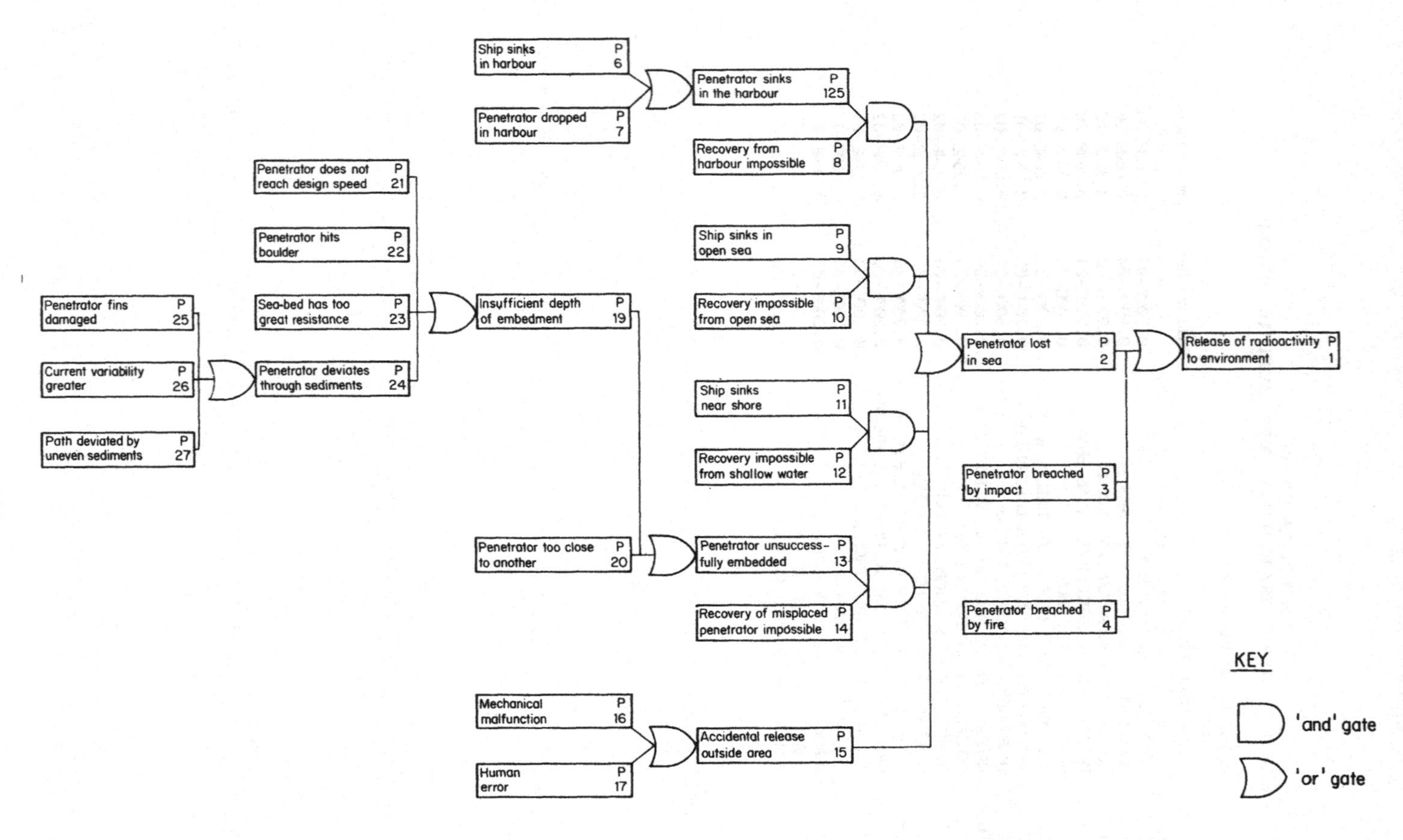

FIG E-1 FAULT TREE ANALYSIS - PENETRATOR METHOD